FLORA OF TROPICAL EAST AFRICA

PASSIFLORACEAE

W. J. J. O. DE WILDE

(Rijksherbarium, Leiden)

Climbers or sometimes erect herbs, mostly provided with axillary tendrils (tribe *Passifloreae*), or erect shrubs or trees (tribe *Paropsieae*), glabrous or hairy, rarely thorny. Leaves mostly alternate, simple or compound, entire or lobed, often with glands on petiole and blade; stipules small, sometimes caducous. Inflorescences axillary, either cymose (*Passifloreae*), ending in 1–several tendrils or not, or racemose (*Paropsieae*); bracts and bracteoles usually small. Flowers hermaphrodite or functionally unisexual (then plants often dioecious or polygamous); stipe articulate to pedicel; hypanthium saucer-shaped to tubiform. Sepals 4–5(–6), imbricate, free or partially connate (*Adenia* in part), often persistent. Petals (3–)4–5(–6), imbricate, rarely absent. Corona extrastaminal, inserted on the hypanthium, various, composed of hairs, or of 1 or more whorls of thread-like processes or scales, or tubiform or cup-shaped, or absent (*Adenia* in part). Disk mostly extrastaminal, annular or composed of 5 mostly strap-shaped parts (*Adenia*), or absent. Stamens 4-many, inserted on the hypanthium or on an androgynophore, if few alternipetalous, free or partially connate; anthers 2-thecous, basifixed to dorsifixed, versatile or not, sometimes apiculate, opening lengthwise. Ovary superior, sessile or on a gynophore or androgynophore, 1-locular, 3–5(–6)-carpellate, with 3–5(–6) parietal placentas; ovules mostly numerous, anatropous; styles 1 or 3–5, very short to distinct, free or partially united; stigma capitate to subglobose, sometimes much divided (*Adenia*). Fruit a loculicidally 3–5-valved capsule, or berry-like. Seeds mostly compressed, enveloped by a membranous or mostly pulpy aril; funicles often distinct; testa crustaceous, mostly pitted; endosperm horny; embryo large, straight, with foliaceous cotyledons.

A pantropical family of about 500 species, comprising 18 genera.

The family is conceived here in the sense of A. P. de Candolle, Prodr. 3: 321 (1828), i.e. including the group of arboreous non-lianescent genera known as tribe *Paropsieae*, as well as the group of generally lianescent genera provided with tendrils, the present tribe *Passifloreae*. Most British authors, from J. D. Hooker, G. P. 1: 810 (1867) onwards, also included the paropsiaceous genera in Passifloraceae.

For various reasons tribe *Paropsieae* was formerly included in Flacourtiaceae by several German authors. Recent investigations have shown that the tribe fits best in Passifloraceae, as pointed out by the author in Blumea 19: 99–104 (1971). Tribe *Paropsieae* is almost confined to Africa, and contains 6 small genera, of which 3 occur in the Flora area.

Tribe *Passifloreae* is composed of 12 genera, distributed over the Old and New Worlds. *Passiflora* is the largest genus with about 370 species, about 350 in America, and about 20 in Australasia; it is not represented in Africa, but two species, *P. foetida* L. and *P. suberosa* L. are locally established introductions, quite often found in ruderal or otherwise disturbed places. Several other *Passiflora* species are introduced as ornamentals (e.g. *P. mollissima* (Kunth) Bailey) or for the edible fruits (*P. edulis* Sims and others), and these occur quite often as garden escapes. Though not indigenous to the Flora, the genus *Passiflora* is entered in the key. Under the genus description a key to

1

the species occurring in the Flora area is given, and the most commonly encountered species are treated fully.

The second largest genus is *Adenia*, mainly represented in Africa and Madagascar, and with comparatively few species in Indo-Malesia. The genus *Basananthe*, with 25 species, is confined to tropical Africa. The remaining genera are monotypic, or contain but a few species, and are all more or less local.

For a review of the family (excluding *Paropsieae*) see Harms in E. & P. Pf., ed. 2, 21: 470–507 (1925). The American representatives of the family are extensively treated by Killip, Field Mus. Nat. Hist. Chicago, Bot. 19: 1–613 (1938); Contr. U. S. Nat. Herb. 35: 1–23 (1960). A treatment of the family for the Malesian area is given by de Wilde in Fl. Males., ser. 1, 7: 405–434 (1972). A monograph of the genus *Adenia*, including many general aspects of the family, is published in Meded. Landbouwhogeschool Wageningen 71–18: 1–281 (1971); an account of the Old World Passifloras in Blumea 20: 227–250 (1972), and of the genus *Basananthe* in Blumea 21: 327–356 (1974). For the delimitation of genera in tribe *Passifloreae* see de Wilde, Blumea 22: 37–50 (1974).

Many members of the family are toxic. Among the toxic constituents are cyanogenic glucosides and for example a toxalbumine, called modeccine, reported from the fruits of *Adenia* (*Modecca*) *digitata*. Other species with highly poisonous fruits are *A. volkensii* and *A. scheffleri*. See also Watt & Breyer-Brandwijk, Medic. & Pois. Pl. S. & E. Afr., ed. 2: 826–830 (1962); Verdcourt & Trump, Common Poisonous Pl. E. Afr.: 37 (1969).

Shrubs or trees, never with tendrils; inflorescences racemose or ? cymose, borne on shoots developed from the
  axillary (not serial-accessory) bud; vegetative ramification through the axillary bud; stamens 5 to many
  (tribe *Paropsieae*):
Style single, with 1 (sometimes faintly lobed) stigma;
  corona double; stamens numerous   .   .   .   1. **Barteria**
Styles (2–)3–4(–6); corona single; stamens 5 or 10–16:
  Sepals (4–)5; stamens 5, ± connate at base   .   .   2. **Paropsia**
  Sepals 4; stamens 10–16, free .   .   .   .   .   3. **Viridivia**
Mostly tendril-bearing climbers or erect herbs (rarely, in
  the New World, shrubs or trees); inflorescences
  cymose, or rarely pseudo-racemose on short-shoots
  developed from the serial, accessory bud; vegetative
  ramification through the accessory bud; stamens
  4–8(–10*) (tribe *Passifloreae*):
Androgynophore distinct, much longer than the ovary;
  flowers hermaphrodite   .   .   .   .   .   4. **Passiflora**
Androgynophore shorter than the ovary or absent;
  flowers hermaphrodite or unisexual:
  Flowers mostly unisexual; corona little developed,
    at most composed of short hairs, or absent; disk
    consisting of 5 strap-shaped parts, or absent   .   5. **Adenia**
  Flowers hermaphrodite or functionally unisexual;
    corona well developed, composed of long threads
    united at base; disk annular or absent:
  Corona double; stamens inserted inside the inner
    corona; disk usually present   .   .   .   6. **Basananthe**
  Corona single; stamens inserted on a very short
    androgynophore, shortly connate at base;
    disk absent:
    Leaves simple, though sometimes deeply incised;
      flowers largely 4-merous; style single,
      stigma (3–)4-lobed .   .   .   .   .   7. **Schlechterina**
    Leaves 3- or 5-foliolate; flowers 5-merous; styles
      3 .   .   .   .   8. **Efulensia**

* Stamens in *Basananthe* rarely up to 9, in *Mitostemma* Mast. (America) 8 or 10.

1. BARTERIA

Hook. f. in J.L.S. 5 : 14, t. 2 (1861) & in G.P. 1 : 812 (1867) ; Gilg in E. & P. Pf., ed. 2, 21 : 415, fig. 164, 183/J-M (1925) ; de Wilde in Blumea 19 : 99–104, fig. 1/h (1971) ; Sleumer in Blumea 22 : 13 (1974)

Shrubs or trees without tendrils ; branchlets usually hollow, at least in part, often inflated, perforated, and inhabited by ants, pubescent or glabrescent. Leaves alternate or subdistichous, not lobed, elliptic to lanceolate, usually leathery, subsessile or petiole short, often shortly winged, decurrent on the branchlets or not ; margin finely crenulate to subentire, set with many small glands ; nerves pinnate ; stipules small, caducous. Inflorescences axillary, sessile, up to 4-flowered fascicles, or horseshoe-shaped, 4–9-flowered, or flowers solitary ; bracts numerous, imbricate, brown. Flowers sessile, hermaphrodite, large, fragrant ; hypanthium thickish, shallowly cup-shaped. Sepals 5, large, imbricate, free. Petals 5, imbricate, resembling the sepals. Corona double, the outer membranous, tubiform, straight or ± folded, with laciniate edge, the inner low, fleshy, with shallowly lobed edge. Stamens many, nearly hypogynous, ± arranged in 2 rows ; filaments united into a tube for the lower one-fourth to half ; anthers oblong-linear, 2-thecous, sub-basifixed. Ovary sessile, placentas 3–4, many-ovuled ; style single, conspicuous, thick, intruding in the stigma ; stigma large, variable in shape, hemi-globose to obtuse-conical, smooth, entire or distally ± (3–)4-lobed, or rarely distinctly 4-lobed, often larger than the ovary. Fruit coriaceous, ? indehiscent but showing (3–)4 valve-sutures, subglobose, subsessile. Seeds many, ovoid, ± flattened, ± 5 mm., arillate ; testa crustaceous, scrobiculate.

A genus now considered to comprise one single variable species, divided into two sub-species, in tropical Africa from S. Nigeria south to Angola (Cabinda), east to Uganda and W. Tanzania.

B. nigritana *Hook. f.* in J.L.S. 5 : 15, t. 2 (1861) ; Mast. in F.T.A. 2 : 510 (1871), as " *nigritiana* " ; Sleumer in Fl. Afr. Centr. : 14 (1974). Type : Nigeria, Niger [Nun] R., *Barter* 2119 (K, holo. !, P, iso.)

Shrub or tree to ± 10 (outside E. Africa up to 25) m., sometimes deciduous ; stem up to 30(–50) cm. in diameter ; bark whitish, rough. Leaf-blades variable in shape, elliptic to oblong or ovate-oblong, or oblanceolate, base rounded to long-cuneate (especially in terminal leaves), top ± obtuse to acute, or acuminate up to 2 cm., 10–30(–45) by 4–10(–19) cm., subcoriaceous, glabrescent or glabrous ; nerves 9–19 pairs, ± prominent on both sides ; petiole 5–10(–15) mm. long, 2–8 mm. wide, mostly shortly winged and decurrent on the branchlets as a low ridge. Inflorescences 1–4-flowered, or outside E. Africa up to 9-flowered ; bracts orbicular to ovate, obtuse or shortly acuminate, 3–15 mm. long, outside ferrugineous pubescent or glabrous, margin ciliate. Flowers large, white. Sepals elliptic to oblong, acuminate, 2·5–4 by 1–1·5 cm., ferrugineous silky or glabrescent outside. Petals resembling the sepals, equal in size or either somewhat smaller or larger, mucronate. Outer corona membranous, ± 1 cm. high ; inner corona a fleshy, shallowly lobed, 3–5 mm. high rim. Stamens ± 3 cm. ; filaments connate for nearly half-way, glabrous ; anthers 3–6 mm. Ovary glabrous, subglobose or with (3–)4–5 bulges in the upper part ; style 10–20 mm. ; stigma conical to subglobose, yellow. Fruit green-yellow to dull orange or reddish, ± 1·5–3 cm. in diameter.

SYN. *B. braunii* Engl. in E.J. 14 : 392 (1891) & 15 : 588 (1893). Type : Cameroun, Grand Batanga, *J. Braun* (B, holo.†, HBG, iso. !)
    B. nigritana Hook. f. var. *uniflora* De Wild. & Th. Dur., Contr. Fl. Congo 1 : 24 (1900). Type : Zaire, Talavanje Forest, *Cabra* (BR, holo. !)

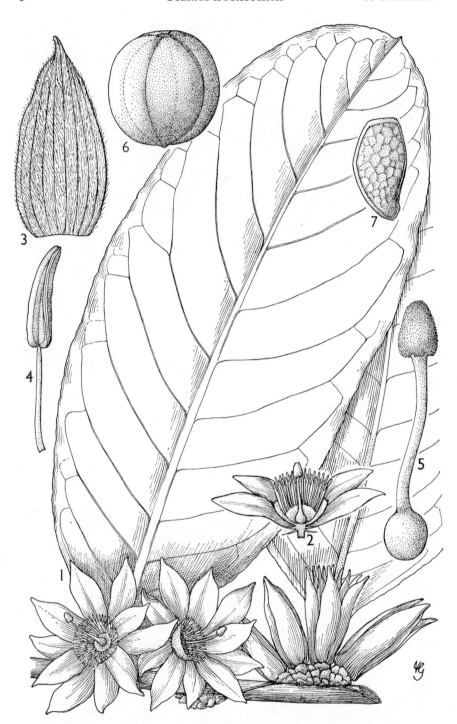

Fig. 1. *BARTERIA NIGRITANA* subsp. *FISTULOSA*—**1,** flowering branchlet, × ⅔; **2,** longitudinal section of flower, × ⅔; **3,** sepal, × 2; **4,** stamen, × 6; **5,** pistil, × 3; **6,** fruit, × 1; **7,** seed, × 4. 1–5, from *Leeuwenberg* 5248, 5250; 6, from *Binuyo* 41445; 7, from *Maitland* 211. Drawn by Victoria Goaman.

subsp. **fistulosa** (*Mast.*) *Sleumer* in Blumea 22 : 14 (1974). Type : Fernando Po, *Mann* 268 (K, holo. !, P, iso. !)

Branchlets usually hollow throughout, cylindrical. Leaves mostly pseudodistichous, variable in shape, in E. Africa usually acuminate, 20–45 by 6–15(–19) cm.; nerves 12–19 pairs; petiole usually flattish, broad, shortly winged, 5–15 by 4–8 mm., decurrent on the branchlets. Inflorescences 1–4-flowered axillary fascicles, or (outside E. Africa) 4–9-flowered in the shape of a horseshoe. Anthers 3–6 mm. long. Fig. 1.

UGANDA. Masaka District : Misozi [Musozi], Jan. 1903, *Bagshawe* 93 ! & Bukakata, 28 Nov. 1942, *A. S. Thomas* 4093 ! ; Mengo District : Kyagwe, 24 Jan. 1951, *Dawkins* 709 !
TANZANIA. Bukoba District : Bukoba, 14 Feb. 1936, *Gillman* 469 ! & Rubare, July 1951, *Eggeling* 6235 !
DISTR. **U**4 ; **T**1 (see further under the note)
HAB. Lake-zone forest, clearings, forest edges ; 1000–1500 m.

SYN. *B. fistulosa* Mast. in F.T.A. 2 : 511 (1871) ; De Wild., Miss. Laurent : 250, t. 91, 92, fig. 33–36 (1906) ; I.T.U., ed. 2 : 312 (1952) ; A. & R. Fernandes in C.F.A. 4 : 216 (1970).
  B. dewevrei De Wild. & Th. Dur., Ann. Mus. Congo, Bot., sér. 2, 1 : 8 (1899) ; De Wild., Miss. Laurent : 247, fig. 32 (1906). Type : Zaire, Bangala, *Dewèvre* 869 (BR, holo. !)
  B. fistulosa Mast. var. *macrophylla* De Wild. & Th. Dur., Reliq. Dewevr. : 98 (1901). Type : Zaire, Mbandaka [Coquilhatville], *Dewèvre* 597 (BR, holo. !)
  B. acuminata Bak. f. in J.L.S. 37 : 155 (1905) ; I.T.U., ed. 2 : 313 (1952). Type : Uganda, Misozi [Musozi], *Bagshawe* 93 (BM, holo. !)
  B. stuhlmannii Engl. & Gilg in E.J. 40 : 479 (1908) ; T.T.C.L. : 448 (1949). Types : Tanzania, Bukoba, *Stuhlmann* 986, 987, 1024, 1025 & 3661 (B, syn.†)
  B. urophylla Mildbr., Z.A.E. 1910–11, 2 : 97 (1922), *nom. nud.*

NOTE. The infraspecific taxonomy of *Barteria* is not quite clear. In a recent study Sleumer (loc. cit.) decided upon the distinction of two subspecies within one single variable species. The subspecies *nigritana* is regarded as a taxon occurring in the coastal areas of W. Africa, of littoral and sublittoral bush and forest, and inner mangrove. It is mainly distinguished by more slender petioles 2–4 mm. wide, smaller leaf-blades 10–25 cm. long, with 9–11(–14) pairs of nerves. The branchlets are solid except for short spindle shaped swellings, the inflorescences axillary, 1–4-flowered, the anthers 3–4 mm. long. In subsp. *fistulosa* the hollow branchlets are apparently always inhabited by large brown stinging ants, whereas in subsp. *nigritana* small blackish ants are found.

2. PAROPSIA

Thouars, Hist. Veg. Isles Austr. Afr. 1 : 59, t. 19 (1805) ; Hook. f. in G.P. 1 : 812 (1867) ; Gilg in E.J. 40 : 470 (1908) & in E. & P. Pf., ed. 2, 21 : 414 (1925) ; Perrier in Fl. Madag., Fam. 143 : 28 (1945) ; Sleumer in Fl. Males., ser. 1, 4 : 645 (1954) & in B.J.B.B. 40 : 49 (1970) ; Presting in Pollen et Spores 7 : 210 (1965)

Trichodia Griff. in Notul. 4 : 570 (1854)
Hounea Baill. in Bull. Soc. Linn. Paris 1 : 301 (1881) ; Gilg in E. & P. Pf., ed. 2, 21 : 413 (1925)

Shrubs or trees, mostly hairy, without tendrils. Leaves alternate, ± distichous, not lobed, ± oblong, petiolate, margin subentire to mostly distinctly serrate-dentate or repand-dentate ; glands at the tips of the teeth, and on the blade margin (rarely on the surface) towards the base ; stipules small, mostly caducous. Inflorescences 1–many-flowered, subfasciculate, axillary, peduncled or not, yellow to reddish-brown hairy, sometimes forming terminal panicles by reduction or abcission of subtending leaves ; bracts linear, caducous. Flowers hermaphrodite, pedicellate, often fragrant. Hypanthium shallowly cup-shaped ; sepals (4–)5, imbricate, persistent. Petals (4–)5, imbricate. Corona composed of hairy threads in 1 series, ± free or connate at base, or these collected into 5 bundles opposite the petals. Stamens 5, opposite the sepals ; filaments glabrous or pilose at base, faintly

connate at base, inserted at base of ovary or on a short gynophore; anthers
oblong to linear, ± sagittate, subdorsifixed. Gynophore short or absent;
ovary subglobose to ovoid, placentas 3–5, many-ovuled; styles (2–)3(–5),
free, sometimes pilose; stigmas subreniform or capitate. Fruit a 3(–5)-valved
capsule, subglobose, sometimes shortly stiped. Seeds ovoid, compressed,
5–7 mm., arillate; testa crustaceous, scrobiculate.

Species 11, in tropical Africa (4), Madagascar (6) and E. Malesia (1).

Sepals 6–10 mm.; ovary after anthesis sparingly pilose or
    glabrous; syles glabrous also at base; flowers normally
    precocious, i.e. present before the deciduous leaves;
    pollen 3-colpate:
    Leaf-blades oblong to elliptic-oblong, rarely subovate-
        elliptic, lower surface soon glabrescent; petiole
        2–3(–4) mm.; corona-threads connate for the lower
        one-third; capsule glabrous.   .   .   .   . 1. *P. guineensis*
    Leaf-blades subovate-elliptic, broadly oblong, or elliptic,
        lower surface long pilose or tomentose; petiole
        5–6 mm.; corona-threads two-thirds connate;
        capsule sparingly and shortly pilose   .   . 2. *P. braunii*
Sepals 17–25 mm.; ovary after anthesis tomentose; styles
    pilose at base; flowers present with the leaves; pollen
    6-colpate   .   .   .   . 3. *P. grewioides*
                                                                  var. *orientalis*

1. **P. guineensis** *Oliv.* in J.L.S. 8: 161 (1865); Mast. in F.T.A. 2: 505
(1871); Gilg in E.J. 40: 471 (1908); Engl., V.E. 3 (2): 572 (1921); I.T.U.,
ed. 2: 313 (1952); F.W.T.A., ed. 2, 1: 201 (1954); A. & R. Fernandes in
C.F.A. 4: 212 (1970); Sleumer in B.J.B.B. 40: 54 (1970). Type: Nigeria,
Old Calabar, *Thomson* 31 (K, holo.!)

Shrub or tree up to 15(–20) m., deciduous; stem up to 50 cm. in diameter;
branches glabrous, when young brownish or reddish-brown tomentose;
lenticels elliptic-oblong, whitish. Leaf-blades elliptic-oblong, rarely
subovate-elliptic, top subacute to subobtuse, sometimes shortly acuminate,
base broadly cuneate to subrotund, chartaceous, ± asymmetrical,
(5–)8–14(–18) by 3–5(–7·5) cm., glabrous above except midrib, pilose beneath,
soon glabrescent except midrib, margin repand-dentate (teeth ± 0·5 mm.);
midrib prominent beneath, lateral nerves 7–8 pairs, ± upward curved,
venation loosely prominent-reticulate; petiole 2–3(–4) mm. Inflorescences
fasciculate, 3–4(–7)-flowered, often arranged into panicles up to 30 cm. long
and wide, in the axils of fallen or very young leaves; branches red-brown
tomentose; bracts ovate, cucullate, caducous. Flowers fragrant; pedicels
slender, 4–9 mm., red-brown tomentose. Sepals subspathulate-lanceolate,
6–7(–10) by 2–3 mm., tomentose outside, glabrous inside. Petals oblong-
subspathulate, 8–10 by 2–3(–4) mm., white, glabrous. Corona 3(–4) mm.
high, composed of threads which are connate in the lower one-third and
glabrous, free parts rather long pilose. Filaments 6–7 mm.; anthers (2–)3
mm. Gynophore 1·5–2 mm.; ovary glabrous; styles slender, glabrous,
± 2 mm. Fruits mostly on nearly bare trees, glabrous, 1·5–2 cm. in diameter.
Fig. 2.

UGANDA. Mengo District: Kyiwaga, 7 Sept. 1949, *Dawkins* 360! & Entebbe, Sept.
    1945, *Eggeling* 5555! & Mubango, Dec. 1914, *Dummer* 1378!
DISTR. U2, 4; west and central Africa, from Ghana to Uganda, south to Angola
HAB. Rain-forest, riverine forest, often as understorey shrub or tree; 1000(–1500) m.

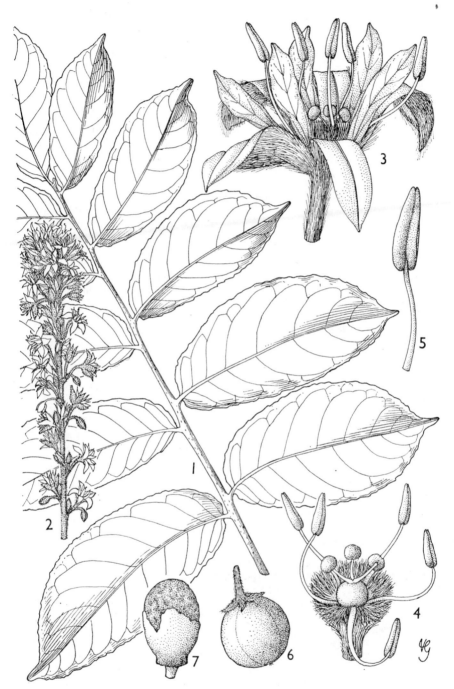

FIG. 2. *PAROPSIA GUINEENSIS*—**1,** leafy branchlet, × ⅔; **2,** part of inflorescence, × ⅔; **3,** flower, × 6; **4,** same, with tepals removed, × 6; **5,** stamen, × 12; **6,** fruit, × 1½; **7,** seed, × 4. 1, from *Dawkins* 360; 2–5, from *Toussaint* 2481; 6, 7, from *Letouzey* 4494. Drawn by Victoria Goaman.

ᵢ SYN.   *Hounea guineensis* (Oliv.) Warb. in E. & P. Pf. III. 6A : 26 (1893)
        Paropsia pritzelii Gilg in E.J. 40: 471 (1908); Engl., V.E. 3 (2): 572 (1921).
        Type: Cameroun, Yaoundé, *Zenker* 727 (B, holo.†, BM, G, K, P, iso. !)

2. P. braunii *Gilg* in E.J. 40: 472 (1908); Engl., V.E. 3(2): 572 (1921); T.T.C.L.: 448 (1949); A. & R. Fernandes in Garcia de Orta 6: 248 (1958); Sleumer in B.J.B.B. 40: 57 (1970). Type: Tanzania, Lindi District, Ruaha–Mtua, *Braun* in *Herb. Amani* 1227 (B, holo.†, EA, iso. !)

Shrubs or trees up to 10 m., much branched, deciduous; branchlets brown tomentose or velutinous. Leaf-blades broadly oblong, elliptic or subovate-elliptic, top subrotund (or shortly and bluntly attenuate), base broadly cuneate or subrotund, asymmetrical, (5–)8–12 by (2·5–)4·5–6(–7) cm., chartaceous, above on midrib and nerves yellowish hairy, scabrous to the touch, beneath pilose (particularly on the nerves) to tomentose, margin inconspicuously dentate or subentire; nerves ± prominent beneath, venation loosely reticulate; lateral nerves 5–7 pairs, upward curved; petiole 5–6 mm. Inflorescences 3–5-flowered fascicles in the axils of fallen leaves, often aggregated into pseudopanicles up to 32 cm. long and wide; bracts cucullate, ± caducous. Flowers precocious, yellowish tomentose, fragrant; pedicels slender, 6–8(–9) mm. Sepals oblong, silky yellow pilose outside, glabrous inside, 6–8 by ± 3·5 mm. Petals oblong, yellowish white, 7–8(–10) by 3 mm., glabrous. Corona in total ± 2 mm. high, composed of threads connate for two-thirds into a glabrous tube, free parts brown pilose. Filaments 6–7 mm.; anthers 2–3 mm. Ovary subglabrous, soon becoming ± pilose; gynophore 1–1·5 mm.; styles slender, glabrous, 1–1·5 mm. Fruits (*Eggeling* 6748) globose, loosely and shortly pilose, ± 1·5 cm. in diameter.

TANZANIA.   Uzaramo District: Pugu Hills, 22 Sept. 1940, *Vaughan* 3143!; Lindi
    District: between Lake Lutamba and Rondo scarp, Nov. 1953, *Eggeling* 6725 ! &
    6726!; Newala District: 19 km. on Newala–Kitangiri road, Nov. 1953, *Eggeling* 6748 !
DISTR.   T6, 8; Mozambique
HAB.   Dry forest and deciduous thicket; 200–400 m.

SYN.   *P. schliebeniana* Sleumer in N.B.G.B. 12: 475 (1935); T.T.C.L.: 448 (1949);
    A. & R. Fernandes in Garcia de Orta 6: 248 (1958). Type: Tanzania, Lindi
    District, Lake Lutamba, *Schlieben* 5442 (B, holo.†, BM, BR, M, P, Z, iso. !)
    P. sp. sensu A. & R. Fernandes in Garcia de Orta 6: 248 (1958)

3. P. grewioides *Mast.* in F.T.A. 2: 505 (1871); Warb. in E. & P. Pf. III.6A: 27 (1893); Engl., V.E. 3(2): 572 (1921); A. & R. Fernandes in Garcia de Orta 6: 653 (1958) & in C.F.A. 4: 213 (1970); Sleumer in B.J.B.B. 40: 63 (1970). Type: Angola, Cuanza Norte, *Welwitsch* 873 (BM, holo. !, C, COI, G, K, LISU, P, iso. !)

Shrub or small treee 4·5–12(–20) m. tall; young branches densely covered with short brown hairs. Leaf-blades usually oblong, sometimes elliptic to obovate-elliptic, top slightly acuminate, subacute, base attenuate, (3·5–)9–11(–13) by (2–)2·5–4(–7) cm., chartaceous to subcoriaceous, glabrous above except sometimes on midrib, appressed hairy only on midrib and sometimes nerves beneath, ultimately glabrescent, margin regularly sub-serrate-crenate (teeth 1–1·5 mm. long, hairy at tips); midrib prominent beneath; lateral nerves 7–14 pairs, straight or curved; venation distinct; petiole slender, 4–8 mm. long, yellowish pubescent. Flowers axillary, solitary or 2–3 in a cluster. Sepals lanceolate-oblong or oblong, (6–)10–25 by 3–10 mm., brownish tomentellous outside, greyish puberulous inside. Petals similar to sepals, but narrower and slightly shorter, greenish yellow or cream. Corona laciniate, 3–4 mm. long, sometimes indistinctly divided into 5 bundles, tomentose outside. Filaments 6–10 mm. long; anthers

linear or linear-oblong, 2·5–3·5 mm. long.   Gynophore 0–1·5 mm. long;
ovary tomentose; styles slender, 2–5 mm. long, tomentose.   Capsule ovoid-
globose to subglobose, 1·2–2·5(–3) by 1·5–2·5 cm., tomentose, sometimes
glabrescent.

var. **orientalis** *Sleumer* in B.J.B.B. 40 : 65 (1970).   Type : Mozambique, Niassa, Mueda
to Mocimboa do Rovuma, *Pedro & Pedrogão* 5312 (COI, holo. !, EA, LMJ, iso. !)

Sepals (17–)20–25 by (4–)5–10 mm.   Fruits ± 1·2 cm. in diameter.

TANZANIA.   Lindi District: Rondo Plateau, Mchinjiri, 1 Oct. 1951, *Bryce* 14 ! & Mar.
1952, *Semsei* 690 ! & Rondo Plateau, 10 Dec. 1955, *Milne-Redhead & Taylor* 7604 !
DISTR.   T8 ; N. Mozambique
HAB.   Dry forest and thicket ; ± 800–850 m.

SYN.   [*P. grewioides* sensu Verdc. in K.B. 11 : 449 (1957) ; A. & R. Fernandes in Garcia
de Orta 6 : 248, t. 1 (1958), *non* Mast. sensu stricto]

NOTE.   Var. *grewioides*, with smaller flowers and larger fruits, occurs in western Africa
from Cameroun to N. Angola, far from the area of var. *orientalis*.

3. **VIRIDIVIA**

Hemsley & Verdc. in Hook., Ic. Pl. 36, t. 3555 (1956)

Shrubs or small trees, yellow to red-brown hairy, without tendrils.   Leaves
alternate, pseudopinnate, deciduous; stipules minute, caducous.   Inflorescences 1–2-flowered, axillary to deciduous leaves at the base of lateral
branches, or arranged in terminal panicles up to 10 cm. across, reddish-brown
hairy.   Bracts small, acute, carinate, caducous.   Flowers hermaphrodite,
pedicellate.   Hypanthium rather narrow, flattish; sepals 4, imbricate,
unequal, ± persistent, 3–7-nerved.   Petals 4, smaller than the sepals,
1-nerved.   Corona extrastaminal, shortly tubiform, irregularly split, hairy
towards the top, and provided with small club-shaped glands.   Stamens
10–16; filaments pilose, inserted on an androgynophore ± 2 mm. long;
anthers oblong, subdorsifixed.   Gynophore ± 4·5 mm.; ovary globose,
placentas 4–5, many-ovuled; styles terminal, 4(–6), free, pilose; stigmas
capitate-reniform.   Fruit a subglobose stipitate 4–5-valved capsule.   Seeds
ovoid, compressed, ± 9 mm., arillate; funicle robust; testa crustaceous,
scrobiculate; embryo large, cotyledons leafy; endosperm not copious.

One species in SW. Tanzania and Zambia.

Closely allied to *Paropsia*, differing mainly by having sepals and petals in fours and
the stamens varying in number between 10 and 16.   In *Viridivia* the inflorescences are
often on short shoots axillary to reduced leaves, in a similar way to that in *Paropsia
guineensis*.

V. suberosa *Hemsley & Verdc.* in Hook., Ic. Pl. 36, t. 3555 (1956).   Type :
Zambia, Kasama District, N. of Lukulu, *Hoyle* 1310 (K, holo. !, FHO, iso.)

Shrub or treelet to ± 8 m., precociously flowering; bark thick, corky with
longitudinal fissures.   Young branches shortly yellow to red-brown hairy.
Leaf-blades ovate or elliptic, rarely elliptic-oblong or lanceolate, 8·5–17(–20)
by 4–7·5(–9·5) cm., top acute to rounded, base broadly cuneate, rounded or
± cordate, margin dentate with bundles of hairs at apex of teeth, both
surfaces (especially beneath) shortly hairy; nerves prominent especially
beneath, lateral nerves 7–11 pairs, reticulation distinct especially on older
leaves; petiole up to ± 5 mm., shortly hairy, with 2 glands at the top.
Inflorescences 1–2-flowered, at base of short-shoots or clustered at the top
of the branches; bracts ovate, caducous, ± 5·5 by 3·5 mm.; pedicels
(4–)7–18 mm., pubescent.   Sepals unequal in size, elliptic to broadly ovate
(top obtuse, base rounded), up to 1·9 by 1·3 cm., hairy outside, glabrous

FIG. 3. *VIRIDIVIA SUBEROSA*—**1**, part of leafy branchlet, × 1; **2, 3**, flowering branchlets, × 1; **4**, longitudinal section of flower, × 3; **5**, fruit, with wall partly removed, × 1; **6**, seed, with part of aril removed, × 2. 1, from *Richards* 5327; 2, from *Bullock* 1347; 3–6, from *Hoyle* 1310. Drawn by Stella Ross-Craig. Reproduced by permission of the Bentham-Moxon Trustees.

inside. Petals narrowly elliptic-spathulate, green-white to creamy yellow, up to 1·8 by 0·6 cm., glabrous, at base contracted into a claw ± 4 mm. long. Corona densely pubescent, ± 3 mm. high. Stamens 10–16; filaments filiform, pilose except towards top, 1·2 cm.; anthers oblong, 2·5–3·5 mm., minutely apiculate. Ovary subglobose, 4–5 mm., densely pubescent; gynophore 4–5 mm.; styles ± 3 mm., stigmas ± 2·5 mm. broad; placentas 3–6-ovulate. Fruit subglobose, ± 3·3 cm. long, 3–4 cm. across, puberulous; gynophore (stipe) ± 1 cm. Seeds ± 9 mm. Fig. 3.

Tanzania. Ufipa District: 24 km. over Tanzania border, 15 Oct. 1962, *Richards* 16831 ! & Lake Tanganyika, Namkala I., 27 Sept. 1964, *Richards* 19102 ! & Kalambo Gorge, 15 Sept. 1959, *Richards* 11454 !
Distr. T4; Zambia
Hab. Deciduous woodland; 750–1100 m.

4. PASSIFLORA

L., Sp. Pl.: 955 (1753) & Gen. Pl., ed. 5: 410 (1754); Harms in E. & P. Pf., ed. 2, 21: 495 (1925); Killip in Field Mus. Nat. Hist. Chicago, Bot. 19: 1–613 (1938); de Wilde in Fl. Males., ser. 1, 7: 407 (1972)

For generic synonyms see Harms, and Killip, loc. cit.

Mostly perennial climbing herbs to large lianas, rarely (not in Africa) shrubs or trees, glabrous or hairy, provided with tendrils. Leaves mostly alternate, unlobed to deeply lobed, palminerved or pinninerved, petiolate; margin mostly dentate, often with small gland-teeth; petiole with or without glands; blade-glands present or not. Stipules minute to large. Inflorescences sessile or peduncled, 1–many-flowered, with or without a simple tendril, or rarely flowers collected into pseudoracemes; bracts and bracteoles small to large, forming a conspicuous involucre or not. Flowers hermaphrodite, 5-merous; hypanthium saucer-shaped to cylindrical. Sepals and petals free, often brightly coloured; petals mostly resembling sepals, membranous, sometimes absent. Corona extrastaminal, variously shaped, simple or mostly composed of a usually complicated outer corona consisting of threads, and flat or plicate inner coronas, sometimes with the addition of a nectary ring or annulus. Androgynophore mostly distinct, 3 mm. or more; stamens 5(–8), free (or in some Asian species partly connate), in older flowers mostly reflexed; anthers dorsifixed, versatile, elliptic to linear. Gynophore absent or sometimes up to 7 mm.; ovary globose to fusiform; styles 3(–4), free or connate at base; stigmas capitate. Fruit usually indehiscent, ± baccate, often with coriaceous exocarp, globose, ellipsoid or rarely fusiform, containing many seeds.

About 370 species, about 350 in the Americas and 20 in SE. continental Asia, Indo-Australia and the west Pacific. The genus is not indigenous in Africa. Species described from Madagascar (*P. calcarata* Mast.) and in the Mascarene Is. (*P. mauritiana* Thouars and *P. mascarensis* Presl) pertain to early introductions from America, the first species most likely being *P. subpeltata* Ortega, the latter two being *P. alata* Dryand.
A number of species is introduced in East Africa as ornamentals (e.g. *P. caerulea* L., *P. mixta* L.f., and *P. mollissima* (Kunth) Bailey) or for the edible fruits with delicate flavour (e.g. *P. edulis* Sims, *P. laurifolia* L., *P. quadrangularis* L.). Two species, *P. foetida* L. and *P. suberosa* L., are locally established weeds in many tropical countries.
For the Flora 22 species are entered in the key; of these 6 of the most commonly encountered species, are treated fully.
The genus is subdivided by Harms (1925) into 21 sections; Killip (1938) accepts 22 subgenera and many sections and series for the American species.

Bracts and bracteoles conspicuous, forming an
   involucre:
   Involucre bracts finely and deeply divided . . 1. *P. foetida* L.
   Involucre bracts not divided:

Hypanthium (calyx-tube) 1·5–9 cm.; involucre
  bracts partially connate, often ± tubiform,
  much shorter than the flower; flowers pink
  or red:
Hypanthium 1·5–2 cm. long .   .   .   .   *P. manicata* (Juss.)
                                                  Pers.

Hypanthium 5–9 cm. long   .   .   .   2. *P. mollissima*
                                                  (Kunth) Bailey

Hypanthium less than 1 cm. long; involucre
  bracts free, or connate at base, then involucre usually ± as large as or larger than
  the flower; flowers of various colours:
Leaves not lobed, pinnately or subpinnately
  nerved:
Stem 4-angular, distinctly winged   .   .   3. *P. quadrangularis*
                                                  L.

Stem not winged:
  Involucre bracts connate at base:
    Stipules foliaceous; involucre bracts
      ± as long as flower:
      Glands on petiole 4–6, thread-like,
        more than 3 mm. long   .   .   *P. ligularis* Juss.
      Glands on petiole 2–4, saucer-shaped,
        stipitate, up to 3 mm. long   .   *P. tiliifolia* L.
    Stipules lanceolate to linear; involucre
      very large, extending beyond the
      flower   .   .   .   •.   .   *P. maliformis* L.
  Involucre bracts free:
    Plant glabrous; leaf margin entire   .   *P. laurifolia* L.
    Plant pubescent; leaf margin serrulate .   *P. serratifolia* L.
Leaves usually lobed or divided, palmately
  nerved:
  Stipules lanceolate or filiform; involucre
    bracts serrate-denticulate:
    Plant glabrous; glands situated at or near
      apex of petiole; flowers whitish,
      4–6 cm. in diameter   .   .   .   4. *P. edulis* Sims
    Plant pilose; glands situated at base of
      petiole; flowers red, ± 10 cm. in
      diameter   .   .   .   .   .   *P. vitifolia* Kunth
  Stipules foliaceous; involucre bracts entire
    or subentire:
    Stipules 1–2 cm. long, falcate, sometimes
      remotely dentate, mucronate; corona
      threads bluish or purplish:
      Leaves (3–)5(–9)-lobed, divided nearly
        to the base   .   .   .   .   *P. caerulea* L.
      Leaves 3-lobed, divided up to ± three-
        fourths of the blade:
        Stem 4-angular (*P. caerulea* × *P.
          quadrangularis*)   .   .   .   *P.* × *allardii*
        Stem terete .   .   .   .   .   *P. violacea* Vell.
    Stipules 1·5–4 cm. long, straight, entire;
      corona threads white   .   .   .   5. *P. subpeltata*
                                                  Ortega

Bracts and bracteoles inconspicuous (filiform or
  linear), not forming an involucre, or foliaceous
  but caducous before anthesis:
  Flowers bright dark red, arranged in pendent
    racemes; involucre bracts caducous before
    anthesis . . . .   *P. racemosa* Brot.
  Flowers of various colours, 1–2(–3) axillary to
    normal leaves; bracts inconspicuous:
    Flowers 1–1·5 cm. in diameter, apetalous, green-
      ish . . . .   6. *P. suberosa* L.
    Flowers more than 1·5 cm. in diameter, with
      petals:
      Leaves 2-lobed, the sinus acute or truncate,
        with a central mucro:
        Leaves pubescent, herbaceous, base cordate   *P. capsularis* L.
        Leaves glabrous, chartaceous or coriaceous:
          Leaf margin not cartilaginous, base
            rounded, not peltate; petiole without
            glands . . . . .   *P. biflora* Lam.
          Leaf cartilaginous at margin; base
            peltate; petiole with 2 glands . .   *P. coriacea* Juss.
      Leaves 3-lobed:
        Leaves variegated, reddish or violet
          beneath; flowers ± 3 cm. in diameter,
          pale greenish . . . . .   *P. trifasciata*
                           Lemaire
        Leaves not variegated, green beneath;
          flowers 4·5–6 cm. in diameter, orange-
          red . . . .   *P. cinnabarina*
                           Lindl.

1. **P. foetida** *L.*, Sp. Pl.: 959 (1753); Killip in Field Mus. Nat. Hist. Chicago,
Bot. 19: 474 (1938). Type: South America (perhaps Lesser Antilles *fide*
Killip), *Linnean Herbarium* 1070.24 (LINN, lecto.)

Climber or creeper to 4 m., annual or biennial, subglabrous or whitish
or yellowish-brown hairy to various degrees, ill-odoured; stem terete. Leaf-
blades entire or usually 3(–5)-lobed up to halfway, suborbicular to ovate in
outline, base cordate, 3–10 by 3–10 cm., 3–5-nerved from near base, usually
membranous; lobes up to 4 cm., usually acute-acuminate, margin entire or
subentire with coarser gland-tipped hairs; petiole 1–6 cm.; stipules subreniform, 0·5–1 cm., deeply cleft into filiform gland-tipped processes. Glands
absent, except gland-tipped stronger hairs or processes on petiole, stipules,
bracts, etc. Inflorescences sessile, 1(–2)-flowered, the straight peduncle
2–6 cm., inserted beside a simple tendril 5–15 cm.; bracts and bracteoles
(1–)2–4 cm., deeply 2–4-pinnatisect, with filiform gland-tipped segments,
forming an involucre just below and enveloping the flower. Flowers ±
3–5 cm. in diameter, pale pinkish or lilac, rarely white. Hypanthium short,
saucer-shaped; sepals ovate-oblong to lanceolate, ± 1·5–2 cm., awned
2–4 mm. dorsally just below the apex. Petals oblong to lanceolate or ±
spathulate, slightly shorter than the sepals. Corona consisting of 2 outer
series of threads ± 1 cm. long, and inward several series of capillary threads
1–2 mm.; operculum membranous, ± erect, denticulate; disk conspicuous,
annular. Androgynophore 4–6 mm.; filaments flattened, ± 5–6 mm.;
anthers 3–5 mm. Ovary globose to ellipsoid, ± 2–3·5 mm., usually glabrous;
styles 4–5 mm. Fruit a rather dry berry, subglobose, 1·5–3 cm. in diameter,

usually glabrous, yellowish to orange, ± enveloped by the persistent involucre. Seeds many, subovoid to ± cuneiform, ± 4–5 mm. long, obscurely 3-dentate at apex.

Kenya. Nairobi, May 1940, ? collector ! (cultivated and escaped)
Tanzania. Lushoto District: Amani, May 1940, *Greenway* 5940 ! & Sept. 1940, *Green-way* 6016 ! (cultivated and escaped); Handeni District: Mgera, July 1960, *Semsei* 3038 !; Uzaramo District: Dar es Salaam, June 1968, *Batty* 126 !; Zanzibar I., Kisim-kazi, Sept. 1960, *Faulkner* 2722 !
Distr. K4; T3, 6, 8; Z; P; America, introduced throughout the tropics
Hab. Sometimes cultivated and often escaped, now common in disturbed places, on coastal sands, etc.; 0–2500 m.

Syn. For synonyms see Killip, loc. cit.

Variation. Most East African specimens can be referred to var. *hispida* (DC.) Gleason (Bull. Torrey Bot. Club 58: 408 (1931); Killip, loc. cit.: 494 (1938)), with glabrous ovary and usually 3-lobed leaves, the whole plant with a rather hispid tomentum. I have refrained from accepting formal varieties as most of those proposed by Killip are not clearly defined.

2. **P. mollissima** (*Kunth*) *Bailey* in Rhodora 18: 156 (1916); Killip in Field Mus. Nat. Hist. Chicago, Bot. 19: 291 (1938); Young in Rec. Auckland Inst. Mus. 7: 159, figs. 29–32 (1970). Type: Colombia, Bogota, *Humboldt & Bonpland* 1767 (P, holo.)

Climber to 20 m., perennial, subglabrous to densely pubescent throughout; stem terete, finely striate. Leaf-blades 3-lobed, the depth varying from about half-way to ± six-sevenths, in outline suborbicular, 5–11 by 6–13 cm., base truncate to cordate, 3(–5)-nerved from the base, membranous to thinly coriaceous, glabrous or subglabrous above, pubescent to various degrees beneath; lobes elliptic to lanceolate, 2·5–6 cm., top acute or up to 1·5 cm. acuminate; margin glandular serrate-dentate up to 2 mm.; petiole 0·7–2·5(–5) cm. Glands on petiole absent or mostly 2–5 pairs, minute, sessile or subsessile; blade-glands, except marginal glands and sometimes a small gland in the lobe-sinuses, absent. Stipules subcircular or obliquely reniform, either ± 2 mm. or ± 0·5(–0·9) cm. in diameter, finely glandular dentate, if small caducous. Inflorescences 1-flowered, the peduncle 1·5–5 cm., inserted beside a simple tendril 5–18 cm.; bracts and bracteoles 2–4 cm., acute-acuminate, connate for half-way to up to ± three-fourths, forming a tubiform involucre. Flowers 5–10 cm. in diameter, pinkish or pinkish red to pinkish orange. Hypanthium tubiform, (5–)6–9 by 0·5–1·5 cm.; sepals oblong, 2–5·5 cm., subobtuse, mucronate below apex. Petals oblong, 2–5 cm., obtuse. Corona a low lobulate edge at the throat of the hypanthium; operculum an inward curved membrane, at the base of the hypanthium. Androgynophore 6–10 cm.; filaments 10–15 mm., dilated; anthers 7–13 mm. Ovary oblong, 10–12 mm., pubescent; styles 10–15 mm. Fruit rather dry, ± ellipsoid, excluding the long gynophore 6–12 cm. long, softly pubescent, yellowish. Seeds many, ellipsoid, ± 6 mm.

Kenya. Naivasha District: Kinangop, 11 July 1965, *Gillet* 16768 !; Kiambu District: Kibata, 20 Aug. 1940, *Greenway* 5987 !; Nairobi, Aug. 1940, *Nattrass* 15 !, 16 ! & 135 ! (cult.)
    Selection of specimens deviating from typical *P. mollissima* (see note). W. Suk District: Kapenguria, Mar. 1956, *Meinertzhagen* !; Kiambu District: Kibata, 20 Aug. 1940, *Greenway* 5985 ! & 27 Aug. 1940, *Nattrass* 39 !; Nairobi, 16 Nov. 1940, *Bell* 45 ! (cult.)
Distr. K2–4; introduced in many tropical countries
Hab. Cultivated as an ornamental in gardens, nurseries, etc., and often escaped, growing in forest edges, clearings, etc.; 1000–3000 m.

Syn. *Tacsonia mollissima* Kunth, Nov. Gen. Sp. 2: 144 (1817)

NOTE. *P. mollissima* is extensively cultivated in the South American Andes, and widely distributed in other tropical mountainous regions, and in the subtropics, as an ornamental or for the flavoured passion fruit ("banana passion fruit"). It commonly occurs as an escape. The species doubtlessly hybridizes with other related species of the subgen. *Tacsonia* (Juss.) Tr. & Planch. (see Killip, 1938; Young, 1970), e.g. with *P. mixta* L.f. and *P. pinnatistipula* Cav. A third very closely related species, *P. psilantha* (Sodiro) Killip, is regarded as a hybrid of *P. mollissima* and *P. tripartita* (Juss.) Poir.

The East African material of *P. mollissima* as described above comprises two rather easily segregated forms. About half of the specimens come closest to true *mollissima*, characterized by the following. Stipules rather large, measuring 5 mm. or more, mostly persistent. Leaves usually densely pilose on both surfaces. Hypanthium tube ± 7–10 cm. long, the tepals 2–3·5 cm. long, hence less than half the hypanthium.

The remainder of the East African material possibly belongs to some undescribed form or is a hybrid with an unknown related species. This is differentiated by the following. Stipules small, ± 2 mm., caducous. Leaves usually glabrous or glabrescent above. Hypanthium tube 5–7 cm., the tepals 3–5·5 cm. long, hence longer than half the hypanthium.

These latter specimens cannot be assigned to the related *P. mixta* mainly because of the distinctly angular stem in that species. *P. psilantha* does not agree because of its very narrow hypanthium tube, and the narrowly lanceolate free bract-portions. A part of the present specimens at EA is determined by Killip (in 1947) as "possibly a horticultural hybrid of perhaps *P. mixta* and *P. pinnatistipula*", but in fact the plants show few intermediate features between these species.

Similar difficulties in identifying material of the same alliance of introduced species in New Zealand is discussed by Young (loc. cit.).

3. P. quadrangularis *L.*, Syst. Nat., ed. 10: 1248 (1759); Killip in Field Mus. Nat. Hist. Chicago, Bot. 19: 335 (1938). Type: Jamacia, *P. Browne* (apparently no specimen in LINN; ? S, holo.)

Climber to ± 15 m., perennial, glabrous throughout; stem stout, 4-angled, winged. Leaf-blades entire, broadly ovate to elliptic, 9–20 by 6–15 cm., base rounded to subcordate, apex abruptly acute or shortly acuminate-mucronate; margin entire, pinninerved, membranous; petiole 2–5 cm. Glands on petiole 3 pairs, wart-like; blade-glands absent. Stipules ovate or elliptic, 2–5 cm. long, acute to mucronate, narrowed at base, entire. Inflorescences 1-flowered, the peduncle (pedicel) 1·5–3 cm., inserted beside a simple tendril 10–20 cm.; bracts and bracteoles subovate, base cordate, apex acute-acuminate, 3–5·5 cm., entire, distinct, forming an involucre. Flowers 8–10 cm. in diameter, whitish or pinkish, corona threads purplish and pinkish banded and mottled. Hypanthium shallowly cup-shaped, ± 1·5 cm. wide. Sepals ovate to ovate-oblong, 3–4 cm., concave, cucullate, corniculate. Petals ovate-oblong to oblong-lanceolate, 3–4·5 cm., obtuse. Corona composed of 5 series, outside 2 series of threads 3–5 cm., inward a series of tubercles and one of short threads, the innermost membranous, lacerate; operculum membranous, 4–6 mm. long, inward curved, denticulate; disk (limen) annular. Androgynophore 12–14 mm., thickened and provided with 2 annuli towards base (trochlea); filaments ± flattened, 7–8 mm.; anthers ± 7 mm. Ovary ellipsoid-oblong, 8–10 mm., glabrous; styles 10–12 mm. Fruit oblong-ovoid, 20–30 cm. long, fleshy with thick rind, terete or 3-grooved. Seeds many, broadly obcordate or suborbicular, 7–10 mm.

UGANDA. Mengo District: Nansana, May 1971, *Katende* 870! (cult. and escaped)
KENYA. Nairobi, Aug. 1940, ? collector 37! (cult.)
TANZANIA. Lushoto District: Amani, May 1940, *Greenway* 5937! (cult.)
DISTR. U4; K4; T3; tropical America, introduced in the tropics of the Old World
HAB. In East Africa ornamental and cultivated for the edible, flavoured fruits, quite often escaped

4. P. edulis *Sims* in Bot. Mag. 45, t. 1989 (1818); Killip in Field Mus. Nat. Hist. Chicago, Bot. 19: 393 (1938). Type: cultivated in Europe, probably originally from Brazil (*fide* Killip, loc. cit.)

Climber to ± 15 m., perennial, glabrous throughout (except ovary); stem sometimes ± angular. Leaf-blades 3-lobed up to three-fourths, rarely unlobed, suborbicular to broadly ovate in outline, 5–11 by 6–12 cm., base acute to cordate, 3-nerved from base, subcoriaceous; lobes elliptic to oblong, up to 8 cm., top acute, shortly acuminate; margin serrate; petiole 1–4 cm. Glands on petiole 2, wart-like, at transition to or up to 0·5 cm. below the blade; blade-glands absent. Stipules lanceolate-linear, ± 1 cm. Inflorescences 1-flowered, the straight peduncle (pedicel) 3–6 cm., inserted beside a simple tendril 5–20 cm.; bracts and bracteoles ovate, acute, 1·5–2 cm., glandular-serrate, forming an involucre. Flowers 4–7 cm. in diameter, white, corona threads purplish towards base. Hypanthium cup-shaped, ± 1 by 1–1·5 cm.; sepals oblong, 2–3 cm., corniculate. Petals oblong, 1–2·5 cm., obtuse. Corona composed of several series of threads, the outer 2 series 0·5–2·5 cm., the inner ones much shorter; operculum membranous, incurved, crenulate-fimbriate; disk (limen) cupuliform, entire or crenulate. Androgynophore 6–8 mm., thickened towards base; filaments subsubulate, 6–8 mm.; anthers 8–10 mm. Ovary subglobose to ellipsoid, 3–5 mm., glabrous or shortly pubescent; styles 10–12 mm. Fruit berry-like, with coriaceous-leathery pericarp, globose to ellipsoid, excluding the 0·5–1·5 cm. long gynophore 4–5 cm. in diameter, glabrous, yellow or purplish. Seeds many, ellipsoid, 5–6 mm.

UGANDA. Mengo District: Sezibwa [Sezibura] Falls, Sept. 1961, *Rose* 207!
KENYA. Kiambu District: Muguga, Oct. 1963, *Greenway* 10899!; Nairobi, May 1946, *Bell* 25!; Embu District: Mt. Kenya Forest, Jan. 1967, *Perdue & Kibuwa* 8354!
TANZANIA. Arusha District: Mt. Meru, Nov. 1969, *Richards* 24644!; Lushoto District: Amani, Sept. 1940, *Greenway* 6012! & 6013! (cult.); Morogoro District: Uluguru Mts., Bondwa, Dec. 1968, *Harris* DSM. 401!
DISTR. U4; K4; T2, 3, 6, 7; widely cultivated
HAB. Often cultivated for the flavoured fruit and escaped in forest edges, thickets and disturbed places; 0–2500 m.

SYN. For synonyms see Killip, loc. cit.

5. **P. subpeltata** *Ortega*, Nov. Rar. Pl. Hort. Matrit. 6: 78 (1798); Killip in Field Mus. Nat. Hist. Chicago, Bot. 19: 436 (1938). Type: Mexico (specimen in MA, *fide* Killip)

Herbaceous creeper or climber to 5 m., perennial, essentially glabrous throughout; stem terete. Leaf-blades 3-lobed to about half-way, suborbicular in outline, 4–10 by 4–11 cm., base rounded, truncate or cordate, subpeltate, (3–)5-nerved from base, herbaceous; lobes elliptic to oblong, up to 5 cm., top obtuse or acutish, ± 1 mm. mucronate; margin entire except for a few gland-teeth in or near the lobe-sinuses; petiole 3–6 cm. Glands on petiole 2–5, scattered or ± paired at about the middle, slender, up to 1 mm. long; blade-glands absent. Stipules ovate-oblong, straight, 1·5–4 cm. long, entire or with a few minute gland-teeth, top mucronulate. Inflorescences 1-flowered, the straight peduncle 3–6 cm., inserted beside a simple tendril 4–12 cm.; bracts and bracteoles ovate or broadly ovate, 1–1·5 cm., acute, entire or with a few minute gland-teeth at base, forming an involucre. Flowers 4–5·5 cm. in diameter, white. Hypanthium broadly cup-shaped, 7–10 mm. wide; sepals oblong, 2–2·5 cm., obtuse, with a subapical horn 0·5–1 cm. Petals oblong, 1·5–2 cm., acutish. Corona composed of 4 or 5 series of threads, those of the outer 2 series 1(–1·5) cm. long, those of inner series 2–6 mm.; operculum subplicate, fimbriate-laciniate for about half or less, and with a fringe of inward curved dentiform processes; disk annular; limen with lobulate edge, erect or ± reflexed at the top, closely surrounding the androgynophore. Androgynophore 10–12 mm.; filaments 5–6 mm., dilated; anthers 5–7 mm. Ovary ellipsoid, ± 4 mm., glabrous; styles

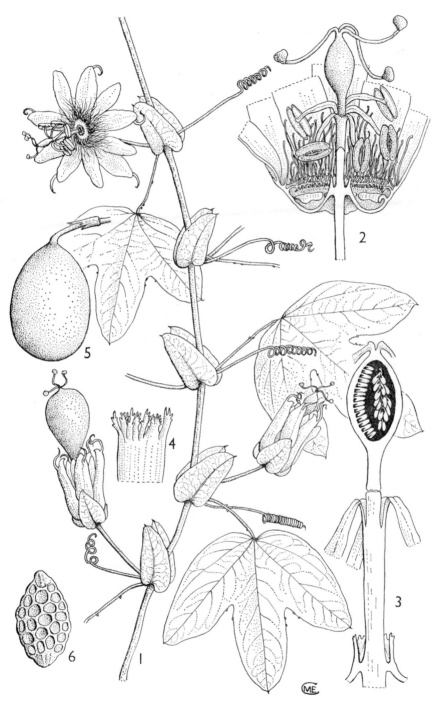

FIG. 4. *PASSIFLORA SUBPELTATA*—**1,** habit, × ⅔; **2,** longitudinal section of flower, × 2; **3,** longitudinal section of androgynophore and ovary, × 4; **4,** part of operculum, outer surface, × 10; **5,** fruit, × ⅔; **6,** seed, × 5. 1–4, from *Greenway* 5819; 6, from *McDonald* in *A.D.* 3271.

8–10 mm. Fruit ± leathery, ellipsoid or subglobose, excluding the 1·5–2 cm. long gynophore 3·5–4·5 cm. long, greenish turning yellow. Seeds many, ellipsoid, 4–5 mm. Fig. 4, p. 17.

KENYA. Kiambu/Nairobi District: Karura Forest, 24 Nov. 1966, *Perdue & Kibuwa* 8089 !; Nairobi, 19 Nov. 1959, *Polhill* 79 !; Kericho District: S. Belgut Reserve, Mar. 1963, *Kerfoot* 4874 !
TANZANIA. Lushoto District: road to Boma, July 1970, *Mshana* 81 ! & Amani, Oct. 1938, *Greenway* 5819 ! (cult.)
DISTR. **K**4, 5 ; **T**3 ; originating from tropical America
HAB. Cultivated as an ornamental ; sometimes freely escaped, e.g. in abandoned plantations, in forest edges, etc. ; 1000–2500 m.

SYN. [*P. eichlerana* sensu Agnew, U.K.W.F.: 166 (1974), *non* Mast.]

NOTE. Killip, in 1947, named the East African material as the similar related species *P. eichlerana* Mast., 1872. This species, according to Killip (1938), differs from true *P. subpeltata* mainly by the operculum which in *P. eichlerana* is laciniate for about half or more, and by the presence of inward curved dentiform processes ; in *P. subpeltata* the operculum is fringed only at the margin, with the entire portion much longer than the teeth or fringe. All the abundant material now available from East Africa, as well as similar material from India, Malesia, Australia, Hawaii, etc., is doubtlessly conspecific. The operculum in all these alien specimens is incised up to about half-way or less, thus rendering the discrimination between *P. eichlerana* and *P. subpeltata* for these specimens difficult. As both species are related in their original country, the introduced plants are possibly a deviating form of one of each, or a hybrid. As *P. subpeltata* is the oldest name I have chosen to accept it in preference to *P. eichlerana*.

6. **P. suberosa** *L.*, Sp. Pl.: 958 (1753); Killip in Field Mus. Nat. Hist. Chicago, Bot. 19: 88 (1938). Type: Hispaniola [Dominica]; specimen grown in Uppsala Botanic Garden, *Linnean Herbarium* 1070.21 (LINN, syn.)

Climber or creeper to 6 m., perennial, glabrous, glabrescent or pubescent to various degrees ; stem ± angular, corky when older. Leaf-blades entire or usually 3-lobed up to four-fifths, subcircular to ovate or oblong in outline, base rounded or cordate, 4–10 by 4–14 cm., 3–5-nerved from base, membranous or subcoriaceous ; lobes triangular to lanceolate, top acute to acuminate ; margin entire ; petiole 0·5–4 cm., with 2 small obconical or wart-like glands at about the middle ; blade-glands usually absent. Stipules linear, 5–8 mm. Inflorescences sessile, 1–2-flowered, with a central tendril, simple, 3–12 cm. ; pedicels 1–2 cm., jointed about half-way ; bracts setaceous, caducous, ± 1 mm. Flowers 1–2 cm. in diameter, pale greenish yellow. Hypanthium saucer-shaped, 3–5 mm. wide ; sepals ovate to lanceolate, subobtuse, 5–10 mm. Petals absent. Corona threads in 2 series, 2–6 mm. ; operculum (inner corona) plicate, minutely fimbriate ; disk annular. Androgynophore 2–4 mm. ; filaments ± subulate, 2–3 mm. ; anthers 1–2 mm. Ovary subglobose-ellipsoid, 1–2 mm., glabrous ; styles 2–3 mm. Fruit a berry, subglobose, 0·8–1·5 cm. in diameter, glabrous, purple-blackish. Seeds several–many, subovoid, 3–4 mm. long.

UGANDA. Mengo District: Entebbe, Mar. 1935, *Chandler* 1144 ! (escaped)
KENYA. Nairobi, Nov. 1940, ? collector ! (escaped)
TANZANIA. Lushoto District: Amani Nursery, Feb. 1939, *Greenway* 5860 ! (cult.)
DISTR. **U**4 ; **K**4 ; **T**3 ; America, introduced throughout the tropics
HAB. Escaped from gardens, roadsides, disturbed shady places ; 0–2500 m.

SYN. For synonyms see Killip, loc. cit.

5. ADENIA

Forssk., Fl. Aegypt.-Arab.: 77 (1775); Engl. in E.J. 14: 374 (1891); Harms in E. & P. Pf., ed. 2, 21: 488 (1925); Liebenberg in Bothalia 3: 513 (1939); Perrier de la Bâthie in Fl. Madag., Fam. 143: 2 (1945); de Wilde in Adansonia, sér. 2, 10: 111 (1970) & in Meded. Landbouwhogeschool Wageningen 71–18: 1–281 (1971)

Modecca Lam., Encycl. Méth. Bot. 4: 208 (1797); Hook. f. in G.P. 1: 813 (1867)

Clemanthus Klotzsch in Peters, Reise Mossamb., Bot. 1: 143 (1861)

Ophiocaulon Hook. f. in G.P. 1: 813 (1867); Harv., Gen. S. Afr. Pl., ed. 2: 121 (1868); Engl. in E.J. 14: 385 (1891)

Keramanthus Hook.f. in Bot. Mag. 102, t. 6271 (1876)

For further synonyms see de Wilde, loc. cit. (1971)

Herbaceous to ± woody perennial climbers with tendrils, sometimes erect herbs or shrublets mostly without tendrils, often with a rootstock or tuber, or a swollen main stem, sometimes thorny or spiny, glabrous or sometimes pubescent. Leaves either simple, entire or lobed, or palmately divided or pseudo-compound; glands (0–)1–2 at the blade-base, at or near the top of the petiole, and with or without glands elsewhere on the lower surface or margin of the blade. Stipules minute, narrowly triangular or reniform. Tendrils axillary. Inflorescences axillary, cymose, the middle (or the first 3) flowers often replaced by tendrils; bracts and bracteoles minute, triangular to subulate. Flowers dioecious or rarely monoecious, bisexual or polygamous, campanulate or urceolate to tubular or infundibuliform, mostly greenish or yellowish, always glabrous; stipe articulate at base. Hypanthium saucer- or cup-shaped, or tubular. Sepals (4–)5(–6), free or partially connate into a calyx-tube, imbricate, persistent. Petals (4–)5(–6), free, included in the calyx, sometimes adnate with the calyx-tube, mostly fimbriate or laciniate. Corona annular, or consisting of 5 cap-shaped parts, or of a laciniate rim or membrane, or composed of hair-like processes, or absent. Disk-glands 5, ligulate or strap-shaped, inserted at or near the base of the hypanthium, alternating with the petals, or absent. Male flowers: stamens (4–)5(–6), hypogynous or perigynous (variably inserted on the hypanthium), free or partially connate into a tube; anthers basifixed, oblong to linear, often apiculate, 2-thecous, opening introrsely to laterally; vestigial ovary minute. Female flowers mostly smaller than ♂, with smaller petals; staminodes ± subulate; ovary superior, shortly stipitate or subsessile, globose to oblong; placentas 3(–5), ovules usually numerous; styles 3(–5), free or partially united, sometimes very short; stigmas mostly subglobose, laciniate, plumose or densely woolly-papillate. Fruit a stipitate 3(–5)-valved capsule; pericarp coriaceous to rather fleshy (and hence fruit ± berry-like), rarely somewhat woody, greenish to yellow or bright red. Seeds ± compressed, with crustaceous pitted testa, enclosed in a membranous to pulpy sometimes juicy aril.

About 93 species in 6 sections; five sections in tropical and South Africa, of which two extend to Asia (sect. *Microblepharis* (Wight & Arn.) Engl. and sect. *Blepharanthes* (Wight & Arn.) Engl.), and one to Madagascar (sect. *Adenia*), two sections (sect. *Ophiocaulon* (Hook. f.) Harms and the monotypic sect. *Paschanthus* (Burch.) Harms) are confined to Africa, and one section (sect. *Erythrocarpus* (Roem.) de Wilde) is confined to Australasia. For the Flora 30 species belonging to four sections.

Ample information on various aspects of the genus, among which the leaf-glands, inflorescences, and flower types in connection with the delimitation of the sections, is given in my monograph, in Meded. Landbouwhogeschool Wageningen 71–18: 1–281 (1971).

Gland at blade-base single, on a median hemispherical
    to spathulate appendage; corona 0 or consisting
    of 5 fleshy cap-shaped parts, or of a fleshy some-
    times sinuate rim, never laciniate or composed

of hairs; disk glands absent; sepals mostly free; stigmas sessile or subsessile (sect. *Ophiocaulon*):

Leaves strictly 3-nerved from base, ovate to oblong; secondary venation ± parallel; sepals in ♂ flowers partially connate . . . 30. *A. tricostata*

Leaves 5-nerved from base, or 3-nerved with an additional 1 or more pairs of main nerves from the midrib, emerging at least 2–20 mm. above its base; blade ovate, broadly ovate or ± 3–5-angular or -lobed:

Blade-glands in or close to the axils of the nerves . . . . 24. *A. bequaertii* subsp. *bequaertii*

Blade-glands not restricted to nerve-axils, sometimes absent:

Corona a ± broad fleshy rim or 5 broad cap-shaped parts; sepals in ♀ flowers 2·5–5 mm.; fruit 0·8–1·5(–1·8) cm.; blade-glands submarginal or absent. . 26. *A. gracilis* subsp. *gracilis*

Corona absent or inconspicuous (fig. 5/6,7); sepals in ♀ flowers (4–)5–9 mm.; fruit (1·5–)2–4·5 cm.; blade-glands variously on the blade, and/or submarginal, or rarely absent:

Venation on lower leaf-surface distinct, ± regularly reticulate with closed areoles; on upper surface this reticulation clearly visible:

Leaves broadly ovate to orbicular, not lobed; nerves neatly arching towards the blade-tip . . . . 29. *A. stolzii*

Leaves broadly ovate, orbicular or ± triangular, often bluntly 3-lobed; nerves ± straight, the upper pair ending in marginal glands which in lobed leaves are the tips of the lobes . 27. *A. gummifera* var. *gummifera*

Venation on lower leaf-surface less distinct and less regularly reticulate, areoles coarser and less regular; on upper surface reticulation only faintly discernible:

Blade-glands ± 2–4, within and bound to the angle of the upper nerves rather remote from the nerve-axils, rarely absent; leaves entire, suborbicular to bluntly 5-angular; upper nerves straight, ending in marginal glands; fruit usually smooth . . . 25. *A. cissampeloides*

Blade-glands 0, or up to 30, scattered or submarginal; leaves entire or lobed, variously shaped; upper nerves straight or arching towards the tip; fruit rugose or finely warty . . 28. *A. reticulata* var. *reticulata*

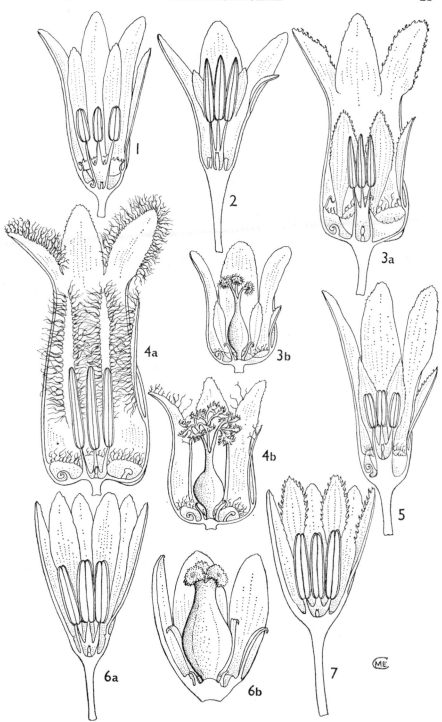

FIG. 5. *Adenia* flowers in longitudinal section. **1**, *A. aculeata* subsp. *aculeata*, ♂, × 3; **2**, *A. globosa* subsp. *globosa*, ♂, × 3; **3**, *A. stenodactyla*, a ♂, b ♀, × 2; **4**, *A. volkensii*, a ♂, b ♀, × 2; **5**, *A. lanceolata* subsp. *scheffleri*, ♂, × 3; **6**, *A. gummifera*, a ♂, × 4, b ♀, × 8; **7**, *A. reticulata*, ♂, × 5. 1, from *Gillett* 12609; 2, from *Faulkner* 1472; 3, from *Milne-Redhead & Taylor* 7789, 7790; 4, from Kew No. 278/46; 5, from *Milne-Redhead & Taylor* 7378; 6a, from *Milne-Redhead & Taylor* 8371; 6b, from *Milne-Redhead & Taylor* 8075; 7, from *Drummond & Hemsley* 4713. Drawn by Mrs. M. E. Church.

Glands at blade-base 0–2, sessile or subsessile, never
   on a single distinct spathulate median appendage
   with narrow insertion; corona absent, or membranous or laciniate, or consisting of hairs; disk
   glands present (in Flora area); sepals free or
   partially connate into a calyx-tube; styles
   distinct:
 Sepals free or nearly so; sepals and petals inserted
    at or at about the same level; anthers wholly
    or largely extending beyond the hypanthium
    (fig. 5/1,2):
  Hypanthium ± as wide as, or wider than long
     (fig. 5/1); corona present, though sometimes
     little developed; flowers ± campanulate;
     the stem prickly or not (sect. *Microblepharis*):
   Male flower (including stipe) more than
      10 mm. long; female flowers 5·5–12 mm.:
    Stem prickly   .   .   .   .   .   1. *A. aculeata*
    Stem unarmed:
       Leaves entire; gland single at blade-
          base   .   .   .   .   .   2. *A. inermis*
       Leaves (3–)5(–7)-lobed; glands 2 at
          blade-base   .   .   .   .   3. *A. racemosa*
    Male and ♀ flowers (including stipe) 3–6 mm.   .   4. *A. wightiana*
                                                              subsp. *africana*

  Hypanthium much longer than wide (fig. 5/2);
     corona absent; flowers tubiform to infundi-
     buliform; stem thorny or not (sect. *Adenia*):
   Stem thorny   .   .   .   .   .   5. *A. globosa*
   Stem unarmed   .   .   .   .   .   6. *A. venenata*
 Sepals partially connate into a calyx-tube which
    extends wholly or partially above the insertion
    of the petals, hence calyx-lobes and petals not
    inserted at the same level, but if approximate
    (because of adnation of the petals to the
    calyx-tube) then stamens (anthers) entirely
    enclosed in the calyx-tube (sect. *Blepharanthes*):
  Leaves shallowly to deeply lobed or entire, not
     digitately compound:
   Plant erect or suberect, herbaceous or shrub-
      like, 0·1–1·5 m. tall, without tendrils:
    Leaves glabrous, with entire margin, oblong
       to lanceolate; calyx-lobes subentire to
       serrulate; anthers 2·5–5 mm.   .   .   23. *A. goetzei*
    Leaves mostly pubescent, with dentate
       margin, or lobed or laciniate, in outline
       ovate to subcircular; calyx-lobes of
       ♂ flowers long woolly fimbriate (fig.
       5/4a); anthers (5–)6–12 mm.:
     Male flowers urceolate, 8–18(–20) mm.
        wide; corona present:
      Leaves entire, dentate, ± peltate;
         glands at blade-base 2, sessile;
         anthers 6–7 mm.   .   .   .   17. *A. keramanthus*
      Leaves 3–7-lobed, very rarely sub-
         entire, not peltate; glands at

blade-base 2, each on a small
auricle; anthers 8–12 mm. .   .   18. *A. volkensii*
Male flowers long-tubular, 20–50 by
3–6 mm.; corona absent or composed
of only a few hairs; leaves lobed,
rarely entire; basal glands 2, sessile
or sometimes on 2 small auricles;
anthers 5–7(–8) mm.   .   .   .   16. *A. ellenbeckii*
Plant 0·5–30 m. long, provided with tendrils:
Corona absent; petals inserted at or above
half-way in the calyx-tube; leaves
pubescent, sometimes only on the
nerves, rarely glabrous; margin
dentate; petals (in ♂ flowers) entirely
long-fimbriate:
Margin of calyx-lobes subentire to
serrulate; anthers reaching to ± the
throat of the calyx-tube .   .   .   15. *A. schliebenii*
Margin of calyx-lobes long and densely
woolly-fimbriate; anthers reaching
to about half-way up the calyx-tube .   16. *A. ellenbeckii*
Corona present, consisting of hair-like
appendages (fig. 5/5), sometimes nearly
absent; petals inserted at the same level
as the corona or up to half-way in the
calyx-tube; leaves glabrous, margin
entire; petals entire, crenulate, serrulate or long-fimbriate:
Glands at blade-base 1–2, sessile or on a
peltate blade-base, or on flattish
auricles which are free or ± connate
over the top of the petiole; hypanthium rather narrow, 3–6 mm. wide,
± as long or longer than wide,
tapering to the base:
Male flowers large, including stipe 20–75
mm.; calyx-tube 5–12 mm. wide;
petals 10–45 mm., long-feathery
fimbriate:
Male flowers (including stipe) 40–75
mm.   .   .   .   .   .   12. *A. dolichosiphon*
Male flowers (including stipe) 20–35
mm.   .   .   .   .   .   13. *A. metriosiphon*
Male flowers smaller, including stipe
13–30 mm.; calyx-tube 3–5(–8)
mm. wide; petals 5–9 mm., entire
or serrulate:
Corona nearly absent; leaf with 2–20
submarginal glands; blade-base
rounded, cordate or hastate   .   14. *A. lindiensis*
Corona consisting of fine hairs; leaf
without submarginal glands;
blade-base attenuate to rounded   19. *A. lanceolata*
Glands at blade-base 2, situated in 2
concave entirely free auricles lateral
at the top of the petiole; hypanthium

much wider than long, 5–15 mm.
wide, ± as wide as the calyx-tube,
base flattish :

Plant slender, up to 5 m. long; stem
terete; leaves elliptic-oblong (2 or
more times as long as wide), base
attenuate to rounded; fruit
fusiform, acute at both ends,
2·5–3·5 by 1·5–2 cm. .   .   .   7. *A. mannii*

Plant robust, up to 30 m. long; older
stems terete, angular or tubercled;
leaves ovate to suborbicular (± 1·5
times as long as wide), base usually
cordate; fruit with rounded top :

Anthers 3·5–5 mm., shorter than the
filaments; ♂ flowers 8–15(–20)
mm. long; fruit subglobose to
ellipsoid, 3·5–5 by 2·5–4 cm.   .   10. *A. schweinfurthii*

Anthers (5–)6–11 mm., much longer
than the filaments; ♂ flowers
10–35 mm. long; fruit globose or
pear-shaped :

Anthers blunt, not apiculate; fruit
globose, ± 3 cm. across;
pericarp 2–3 mm. thick; stem
relatively slender, terete   .   8. *A. panduriformis*

Anthers bluntish to acute, 0·5–2
mm. apiculate; fruit pear-
shaped, 3·5–8 by (1·5–)2·5–4·5
cm.; pericarp 4–10 mm. thick .   9. *A. rumicifolia*

Leaves digitately dissected to the base, the lobes
or leaflets mostly ± stalked :

Hypanthium (at least of ♂ flowers) 6–12 mm.
wide (fig. 5/3) .   .   .   .   .   11. *A. stenodactyla*

Hypanthium or basal part of calyx narrow, up
to ± 5 mm. wide :

Anthers ± curved, connate at apex, 3–5(–6)
mm. long (rarely anthers ± free, but
then less than 4 mm. long); filaments
connate ± half-way or more; corona
usually present .   .   .   .   20. *A. digitata*

Anthers straight, free (even in bud), 4–6
mm.; filaments connate less than half-
way; corona absent or of sparse hairs :

Stipe much longer than the proper ♂
flower; corona absent .   .   .   21. *A. kirkii*

Stipe much shorter than the proper ♂
flower; corona present .   .   .   22. *A. trisecta*

1. **A. aculeata** (*Hook.f.*) *Engl.* in E.J. 14: 375 (1891) & V.E. 1(1): 176,
fig. 144 (1910) & 3(2): 603, fig. 268 (1921); Harms in E. & P. Pf., ed. 2, 21:
491 (1925); de Wilde in Med. Landbouwhogeschool Wageningen 71–18:
65 (1971), excl. subsp. *inermis*, & in Acta Bot. Neerl. 21: 563, fig. 1/a–d,
3/d–e (1972). Type: Somali Republic (S.), Mogadiscio, *Kirk* (K, holo. !)

Climber up to 20 m. long, up to 10 cm. thick at base, without a tuber;

stems (except young parts) strongly prickly, leafless during the greater part of the year; prickles (0·2–)0·5–2 cm., acute, simple, or antler-like (up to twice forked), ± arranged in 4 or 5 rows. Leaf-blades often scabrous, often grey-glaucous, punctate or not beneath, 3–7-lobed up to half-way, sub-orbicular to ovate in outline, base cordate, top obtuse to subacute, 1–7(–16) by 1–7·5(–11) cm., 5-nerved from base; margin entire; lobes up to 3 cm.; petiole 0·5–5 cm. Gland at blade-base 1, situated partly on the shortly peltate blade-base, partly on the top of the petiole; blade glands 0–4(–6). Stipules ± broadly triangular, acuminate, sometimes serrulate, 1–1·5 mm. Inflorescences small, sessile, subsessile or rarely peduncled, grouped in small fascicles axillary to normal leaves (or scars) or mostly in fascicles axillary to much reduced wart-like leaves along short-shoots up to 10 cm., 2–6(–40)-flowered in ♂, 1–3-flowered in ♀, without tendrils; sterile tendrils simple, 5–10 cm. Male flowers narrowly campanulate, including the 1·5–4 mm. long stipe 11–18 mm.; hypanthium cup-shaped, 1·5–3 by 2–4(–5) mm.; calyx-tube absent, sepals lanceolate, obtuse, 7–12 mm., subentire; petals lanceolate, acute to obtuse, 7–9 mm., 5-nerved, finely serrulate; filaments 4–5·5 mm., connate for 2–3·5 mm., inserted at the base of the hypanthium; anthers 3–4 mm., obtuse, up to 2 mm. apiculate; septa ± 2 mm. high; corona membranous, 0·3–1 mm., with laciniate edge; disk glands 0·2–0·4 mm. Female flowers narrowly campanulate, including the 0–1 mm. long stipe 8–12 mm.; hypanthium broadly cup-shaped, ± 1 by 1·5–3 mm.; calyx-tube absent; petals ovate to oblong, ± 4 mm., 3–5-nerved, subentire; staminodes 2–3·5 mm., connate at base; septa 0·5–0·8 mm. high; corona membranous, 0·1–0·5 mm., with sinuate-laciniate edge; disk glands ± 0·2 mm.; gynophore 0·5–1 mm.; ovary broadly ellipsoid, (2–)3–4 mm.; styles connate for ± 0·5 mm., style-arms 0·5–1·5 mm.; stigmas reniform, papillate, each 1·5–2 mm. in diameter. Fruits 1–2 per inflorescence, broadly ellipsoid, distinctly apiculate, excluding the ± 1 mm. long gynophore 1–1·5 by 0·8–1·3 cm.; pericarp thinly coriaceous. Seeds 10–15 per capsule, suborbicular, 3·5–4 mm.

NOTE. Two subspecies are recognized; both are conspicuously prickly. These prickles are emergences from the stem, not comparable with true thorns homologous with inflorescences, as found in *Adenia globosa* or the South African *A. spinosa* Burtt Davy.

subsp. **aculeata**

Prickles of stem 0·5–1 cm., with much broadened 2–5 mm. wide base. Leaf-blades suborbicular to ovate in outline, 3–7-lobed, 2–7(–16) by 2–7·5(–11) cm., mostly distinctly scabrous with fine whitish protuberances on the smaller veins, not punctate. Blade-glands submarginal. Inflorescences grouped in fascicles along short-shoots 1–10 cm., rarely pedunculate along normal shoots. Fig. 5/1, p. 21.

KENYA. Northern Frontier Province: Dandu, 21 Mar. 1952, *Gillett* 12609 !
DISTR. **K**1; E. Ethiopia, Somali Republic
HAB. Deciduous bushland; 500–750 m.

SYN. *Modecca aculeata* Hook. f. in Hook., Ic. Pl. 14: 11, t.1317 (1880)

subsp. **manganiana** (*Chiov.*) *de Wilde* in Meded. Landbouwhogeschool Wageningen 71–18: 68 (1971). Type: Somali Republic, Kisimaio District, *Mangano* (FI, holo. !)

Prickles of stem 0·7–2 cm., at base 1–2(–3) mm. wide. Leaf-blades suborbicular in outline, 5-lobed, 1–5 by 1–5·5 cm., not scabrous, densely purplish brown punctate beneath. Blade glands absent. Inflorescences in fascicles, along up to ± 5 cm. long short-shoots. Flowers and fruit not known.

KENYA. Lamu District: Kiunga, 21 Nov. 1947, *Bally* 5951 !
DISTR. **K**7; southern Somali Republic
HAB. Coastal scrub; 0–100 m.

SYN. *A. manganiana* Chiov., Result. Sci. Miss. Stef.-Paoli, Coll. Bot., Append.: 212 (1916)

2. **A. inermis** (*de Wilde*) *de Wilde* in Acta Bot. Neerl. 21 : 563, fig. 1/e–h (1972). Type: Ethiopia, Bale Province, SE. of Ginner, *J.J.F.E. de Wilde* 7321 (WAG, holo. !)

Climber to ± 8 m.; stem slender, not prickly, growing from a sub-terranean tuber; plant leafless during the greater part of the year. Leaf-blades not scabrous, not punctate beneath, not lobed, ovate-elliptic, base broadly rounded or truncate, subpeltate, top rounded to obtuse, 3–6 by 2–4 cm., 3(–5)-nerved from base; margin entire; petiole 1–2 cm. Gland at blade-base 1, at the transition of blade to petiole; blade-glands absent. Stipules triangular, ± 1 mm. Inflorescences few-flowered, arranged in fascicles axillary to normal leaves or tendrils, not in special inflorescence-bearing short-shoots; fascicles (2–)5–20-flowered in ♂, 1–8-flowered in ♀, without tendrils; sterile tendrils simple, 5–10 cm. Male flowers campanulate, including the ± 1·5–2 mm. long stipe ± 10–11 mm.; hypanthium cup-shaped, ± 1·5 by 2–3·5 mm.; calyx-tube absent, sepals oblong-lanceolate, subobtuse, 6·5–9 mm., subentire; petals lanceolate, subacute, subentire, 5–8 mm., 3–5-nerved; filaments ± 4 mm., connate for ± 2 mm., inserted at the base of the hypanthium; anthers ± 3·5 mm., obtuse; septa ± 0·7 mm. high, corona membranous, ± 1 mm. high, with laciniate-emarginate edge; disk glands ± 0·5 mm. Female flowers campanulate, including the ± 0·5 mm. long stipe 5–6 mm.; hypanthium broadly cup-shaped, ± 1 by 2 mm.; calyx-tube absent, sepals oblong-lanceolate, 4–5 mm.; petals ovate to oblong, subacute, 1·5–4 mm., 3(–5)-nerved, subentire; staminodes ± 2 mm.; gynophore ± 0·5 mm.; ovary subglobose, ± 2 mm.; styles connate for ± 0·5 mm., style-arms ± 1 mm.; stigmas ± reniform, papillate, each ± 1–1·5 mm. in diameter. Fruits 1–4 per fascicle (1–2 per inflorescence), subglobose, not apiculate, excluding the ± 1 mm. long gynophore 1–1·5 by 0·8–1·3 cm.; pericarp thinly coriaceous. Seeds ± 8–10 per capsule, suborbicular, 4–5 mm. in diameter.

KENYA. Northern Frontier Province: Moyale, 1 Mar. 1963, *Bally* 12660 in part! & 19 July 1952, *Gillett* 13611 ! & 8 Aug. 1952, *Gillett* 13698!
DISTR. **K**1; SE. Ethiopia
HAB. Dry montane scrub; 1000–1100 m.

SYN. *A. aculeata* (Hook. f.) Engl. subsp. *inermis* de Wilde in Meded. Landbouwhogeschool Wageningen 71–18 : 69 (1971)

3 **A. racemosa** *de Wilde* in Meded. Landbouwhogeschool Wageningen 71–18 : 64 (1971). Type: Tanzania, Mpwapwa District, Kongwa Forest Research Station, *Wigg* in *E.A.H.* 13734 (EA, holo. !)

Climber to 8 m., glabrous. Leaf-blades dull glaucous green, not punctate beneath, (3–)5(–7)-lobed, suborbicular to broadly ovate in outline, base cordate to truncate, top acute to obtuse, 3–10(–18) by 3–10(–20) cm., 5-nerved from base; margin entire; lobes triangular to elliptic, up to 5 cm.; petiole 1·5–5·5(–7·5) cm. Glands at blade-base 2, on 2 small auricles at the blade-margin at the transition to the petiole; blade-glands (0–)2–4, sub-marginal. Stipules broadly triangular, acute, ± 1 mm. Inflorescences in the axils of much reduced tricuspid leaves ± 2 mm. long, arranged on short-shoots 1–3 cm. long, peduncled for up to 4 mm., 3–7-flowered in ♂, 1–3-flowered in ♀, tendrils 0; sterile tendrils simple, 8–12 cm. Male flowers narrowly campanulate, including the 5–7 mm. long stipe 12–19 mm.; hypanthium cup-shaped, 2–3 by 2–2·5 mm.; calyx-tube 0, sepals lanceolate-linear, subobtuse to subacute, 6–9 mm., entire; petals lanceolate, subacute, 5·5–9 mm., (3–)5-nerved, subentire; filaments (2·5–)3–5 mm., connate for (0·5–)1–2 mm., inserted at the base of the hypanthium; anthers 3–4 mm., subobtuse, apiculate; septa absent; corona composed of a few thick 'hairs'

± 0·5 mm., and of 5 membranous appendages 1 mm., opposite the petals; disk glands ± 0·2 mm. Female flowers campanulate, including the 1–1·5 mm. long stipe 7–8 mm.; hypanthium cup-shaped, 1–1·5 by 2·5 mm.; calyx-tube absent, sepals oblong-lanceolate, obtuse, ± 5 mm.; petals oblong-lanceolate, acute, ± 3 mm., subentire; staminodes ± 2 mm.; septa 0; corona a ± lobed membrane or rim ± 0·2 mm.; disk glands ± 0·2 mm.; gynophore ± 0·5 mm.; ovary ovoid, ± 3 mm.; styles free, ± 2 mm.; stigmas subglobose, papillate, each ± 1·5 mm. in diameter. Fruits 1 per inflorescence, up to 5 on a short-shoot, ovoid-ellipsoid, excluding the ± 2·5 mm. long gynophore 2–3 by 1·5–2 cm.; pericarp ± woody-coriaceous, red. Seeds ± 20 per capsule, broadly ovate, ± 6·5 mm.

TANZANIA. Mpwapwa District: Kongwa Research Station, 5 May 1967, *Matimva* 74! & Jan. 1967, *Wigg* in *E.A.H.* 13734!; Iringa District: Kimiri'matonga Mt., 4 Mar. 1970, *Greenway & Kanuri* 14024!
DISTR. **T**5, 7; not known elsewhere
HAB. Secondary bushland; ± 1000 m.

NOTE. The inflorescence-bearing short-shoots develop from the serial bud, in the axils of sterile tendrils; similar short-shoots are found for example in *A. venenata* and *A. globosa*.

4. **A. wightiana** (*Wight & Arn.*) *Engl.* in E.J. 14: 376 (1891); Harms in E. & P. Pf. III. 6A: 84 (1893) & ed. 2, 21: 492 (1925); de Wilde in Meded. Landbouwhogeschool Wageningen 71–18: 79, fig. 8 (1971) & in U.K.W.F.: 166 (1974). Type: India, Madras, *Wallich* 6764 (K-WALL, holo.!)

Climber to ± 8 m., growing from a tuberous rootstock. Leaf-blades grey-green, entire to deeply 3–5-lobed, broadly ovate to oblong-lanceolate or subtriangular, base cuneate to cordate or truncate, top acute to obtuse, 2–12(–14) by 1·5–11 cm., 3–5-nerved from base and 2–6 nerves from midrib, margin entire to irregularly sinuate or dentate; petiole 0·5–6 cm. Gland at blade-base 1, on the ± fleshy semicircular subpeltate blade-base; blade glands absent, marginal glands (0–)1–3(–4) at either side of the blade. Stipules triangular to broadly reniform, ± 1 mm., deeply dissected with minute glands at the tips. Inflorescences peduncled for (0·5–)2–15 cm., rarely subsessile on short-shoots, cincinnal, up to 30-flowered in ♂, 2–6-flowered in ♀, tendrils 1–3, 0·5–1·5 cm.; sterile tendrils simple or 3-fid, up to 15 cm. Dioecious or monoecious. Male flowers campanulate, including the 1·5–3 mm. long stipe 3–6 mm.; hypanthium cup-shaped, ± 1 by 2–3 mm.; calyx-tube absent, sepals ovate, obtuse to subacute, 1–2 mm.; petals ovate to elliptic-oblong, 1–1·5 mm., 1(–3)-nerved, subentire; filaments 1·5–2·5 mm., connate into a narrow tube for 1·2–1·5 mm., inserted at the base of the hypanthium; anthers ± 0·5 mm., obtuse; septa ± 0; corona consisting of threads 0·5–0·8 mm.; disk glands ± 0·3 mm. Female flowers broadly campanulate, fleshy, including the 0·5–2 mm. long stipe 2·5–6 mm.; hypanthium cup- to saucer-shaped, ± fleshy, 1–2 by 1·5–2 mm.; calyx-tube 0, sepals subovate, obtusish, 1–2 mm., entire; petals obovate-oblong, obtuse, 1–1·5 mm.; gynophore 0·2–0·5 mm., ovary ellipsoid to subglobose, 1·7–2·5 mm.; styles up to 0·5 mm., ± free; stigmas subreniform, papillate-laciniate, each 1–1·5 mm. in diameter. Fruits 1–2 per inflorescence, subglobose to ellipsoid to ovoid, excluding the 1–3 mm. long gynophore 1·5–3 by 1·2–2 cm.; pericarp chartaceous to coriaceous, bright red. Seeds 10–25 per capsule, subovate, 5–6 mm.

SYN. *Modecca wightiana* Wight & Arn., Prodr. Fl. Pen. Ind. Or. 1: 353 (1834); Wight, Ic. Pl. Ind. Or. 1, t.179 (1839); Mast. in Fl. Brit. Ind. 2: 601 (1879) subsp. **africana** *de Wilde* in Blumea 17: 179 (1969) & in Meded. Landbouwhogeschool Wageningen 71–18: 82 (1971). Type: Tanzania, Kondoa District, Kolo, *Polhill & Paulo* 1146 (K, holo. !, BR, EA, LISC, P, iso. !)

Leaf-blades mostly not punctate, entire or shallowly 3–5-lobed, mostly irregularly sinuate-dentate, margin mostly with scattered teeth ± 0·5 mm., and with 0–3(–4) small glands at either side. Stipules broadly reniform, deeply and coarsely dissected with minute glands at the tips. Inflorescences with 3 tendrils, sterile tendrils 3-fid, rarely simple. Male flowers including the 1·5–2 mm. long stipe 3–5 mm.; hypanthium ± 1 mm., sepals 1·5–2 mm.; petals 1–1·5 mm.; anthers ± 0·5 mm. Female flowers broadly campanulate, including the 0·5–1 mm. long stipe ± 3 mm.; hypanthium thickly fleshy, broadly cup- to saucer-shaped, ± 1 by 2–3 mm.; sepals 1–1·5 mm.; petals 1–1·5 mm.

KENYA. Fort Hall District: Thika–Kitui road, 11 Jan. 1972, *Gillett* 19448 !; Machakos District: Chyulu foothills, 25 Apr. 1938, *Bally* 8002 !; Teita District: Maungu Hills, Dec. 1970, *Archer* in *E.A.H.* 14782 !

TANZANIA. Mbulu District: Lake Manyara National Park, 3 Mar. 1964, *Greenway & Kanuri* 11293 !; Lushoto District: Segoma Forest, 23 July 1966, *Faulkner* 3824 !; Morogoro District: Morningside, 23 Mar. 1953, *Drummond & Hemsley* 1747 !

DISTR. **K**4, 6, 7; **T**1–3, 5, 6; not known elsewhere

HAB. Forest edges, deciduous and secondary bushland, wooded grassland; 450–1650 m.

NOTE. Subsp. *wightiana* occurs in S. India and Ceylon, and differs by a number of small characters.

5. **A. globosa** *Engl.* in E.J. 14: 382, fig. 8 (1891); Harms in P.O.A. C: 281 (1895) & in E.J. 24: 168 (1897); Volkens, Kilimandscharo: 18, fig. (1897); Engl., V.E. 1(1): 252, fig. 219 (1910) & 3(2): 595, 607–609, fig. 272 (1921); Harms in Z.A.E. 2: 572 (1913) & in E. & P. Pf., ed. 2, 21: 494, fig. 228 (1925); Verdc. in Bol. Soc. Brot., sér. 2, 38: 97, t.7 (1964); Hutch., Evol. & Phyl. Fl. Pl.: 219, fig. 189 (1969); de Wilde in Meded. Landbouwhogeschool Wageningen 71–18: 128, fig. 17 (1971) & in U.K.W.F.: 165 (1974). Type: Kenya, Teita District, Duruma to Teita, *Hildebrandt* 2858 (B, holo.†)

Shrubs or climbers up to 8 m.; stems erect, curved or scandent, ± succulent, thorned, emerging from a succulent lumpy trunk up to 2·5 m. wide. Leaves caducous, grey-green; blade entire or shallowly 3-lobed, subtriangular to rhomboid, or hastate, base rounded, top acute, 0·3–0·7 by 0·15–0·9 cm., 3(–5)-nerved from base; margin entire; petiole 0·1–0·15 cm. Gland at blade base 1, situated medially on the shortly peltate blade-base; no other glands present. Stipules narrowly triangular, acute, ± 0·5 mm. Thorns in the axils of leaves or leaf-scars, 0·5–8 cm., tip woody, acute; when young bearing half-way 2 subopposite bracts ± 0·5 mm. Inflorescences in the axils of much reduced leaves arranged on short-shoots 0·2–2(–5) cm., situated terminally on the branches in the axils of thorns or rarely on the thorns themselves, peduncled up to 1·5 mm.; 1–5-flowered in ♂, 1–3-flowered in ♀; tendrils 0. Male flowers tubiform to infundibuliform, including the 6–10(–14) mm. long stipe 19–30(–35) mm.; hypanthium tubiform, 5–12 by 2–5 mm.; calyx-tube 0, sepals lanceolate, obtuse, 6–9 mm., entire; petals obovate to ovate or oblong-lanceolate, obtuse or subacute, 4–7·5 mm., 3–5-nerved, denticulate; filaments 2–7 mm., connate for 0–1(–2) mm., inserted up to 7 mm. above the base of the hypanthium; anthers 4–8 mm., obtuse or acute; septa 0(–1) mm.; corona 0; disk glands 0·5–2 mm., sometimes ± connate. Female flowers ± campanulate, including the 1–2 mm. long stipe (6–)8–12 mm.; hypanthium cup-shaped, (1–)1·5–4 by 2·5–4 mm.; calyx-tube 0, sepals lanceolate, obtuse, 5–8 mm., denticulate; petals oblonglanceolate or lanceolate, subacute, (1·5–)2–3 mm., 1-nerved, subserrulate; staminodes 1·5–3·5 mm.; septa 0; corona 0; disk glands 0·2–1 mm.; gynophore ± 1 mm.; ovary ovoid-ellipsoid, 2·5–4·5 mm., ovules 6-30; styles connate for 0·5–1·5 mm., style-arms 1–1·5 mm.; stigmas reniform, papillate, each ± 2 mm. in diameter. Fruits 1 per inflorescence, up to 10 on a short-shoot, subglobose to ovoid-ellipsoid, excluding the 1–3 mm. long gynophore 1·2–2·8 by 1–2 cm.; pericarp coriaceous. Seeds 3–25 per capsule (2–10 ovules per placenta), broadly ovoid, ± 7 mm.

NOTE. The thorns develop in the axils of the leaves, and are homologous with a tendril or a tendril-bearing inflorescence as in other *Adenias*.

Three allopatric subspecies are recognized; according to Verdcourt, loc. cit., the origin of the three taxa is possibly linked with tertiary volcanic activity.

KEY TO INFRASPECIFIC TAXA

Main branches scandent, strongly curved or prostrate and twisted; thorns (1–)2–8 cm., ± as long as or longer than the internodes; inflorescences grouped in the axils of the thorns, scattered along the branches; stamens inserted up to 3 mm. above the base of the hypanthium; anthers 6–8 mm.; ovules 2(–3) per placenta, 6–8 per ovary; fruit 1·2–1·8 cm., containing 3–6 seeds; gynophore 2–3 mm. . . . . . subsp. **globosa**

Main branches erect, curved or subscandent; thorns 0·5–2·5 cm., ± as long as or shorter than the internodes; inflorescences on the apical part of the branches; stamens inserted in the hypanthium 2–7 mm. above the base; anthers 4–6·5 mm.; ovules 2–9 per placenta, 7–30 per ovary; fruit 2–2·8 cm., containing 7–25 seeds; gynophore 1–2 mm.:

Main branches erect or but little curved; stamens inserted 3·5–7 mm. above the base of the hypanthium; ovules 5–9 per placenta, 18–30 per ovary; fruit containing 15–25 seeds . . subsp. **pseudoglobosa**

Main branches strongly curved or subscandent; stamens inserted 2–3(–4) mm. above the base of the hypanthium; ovules 2–5 per placenta, 7–12 per ovary; fruit containing 7–12 seeds . subsp. **curvata**

subsp. **globosa**; de Wilde in Meded. Landbouwhogeschool Wageningen 71–18: 130 (1971) & in U.K.W.F.: 165 (1974)

Main branches up to 8 m., scandent or strongly arching or prostrate and twisted. Thorns strong, (1–)2–8 cm., strictly patent, ± as long as or longer than the internodes. Inflorescences on short-shoots in the axils of the thorns, scattered along the branches, not terminal. Male flowers including the 7–11 mm. long stipe 19–30 mm.; hypanthium 5–9 mm.; sepals 7–9 mm.; petals 4–7·5 by 1·5–2·5 mm.; filaments inserted in the hypanthium up to 3 mm. above the base, 3·5–7 mm., connate for up to 2 mm.; anthers 6–8 mm.; disk glands 0·7–2 mm. Female flowers including the 1–2 mm. long stipe (6–)8–12 mm.; hypanthium (1–)1·5–2 mm.; sepals (4–)5–8 mm.; petals 1·5–2·5 by 0·7–1 mm.; staminodes inserted at the base of the hypanthium, (1·5–)2–3 mm.; disk glands (0·2–)0·7–1 mm.; pistil (4–)5–9 mm.; ovary 2–4 by 1·7–3 mm.; ovules 2(–3) per placenta. Fruit excluding the 2–3 mm. long gynophore 1·2–1·8 cm. Seeds 3–6 per capsule. Fig. 5/2, p. 21.

KENYA. Machakos District: Makindu, 29 Aug. 1959, *Verdcourt* 2369!; Kitui District: Mutomo Hill, 28 May 1965, *Bally* 12797!; Kwale District: Samburu–Mackinnon Road, 30 Aug. 1953, *Drummond & Hemsley* 4053!
TANZANIA. Moshi District: Chala Lake, *Gilbert* 60!; Pare District: Maramba–Kihurio, 17 June 1942, *Greenway* 6474!; Handeni District: Korogwe–Handeni, 25 July 1954, *Faulkner* 1472!
DISTR. **K**4, 7; **T**2, 3; southern Somali Republic
HAB. Deciduous and dry evergreen bushland, often in rocky places, sometimes on sandy or black clay soils; 100–1500 m.

subsp. **pseudoglobosa** (*Verdc.*) *de Wilde* in Blumea 17 : 180 (1969) & in Meded. Land⁻ bouwhogeschool Wageningen 71–18 : 132 (1971) & in U.K.W.F. 165 (1974). Type : Kenya, Kiambu District, Ndeiya Grazing Scheme, *Verdcourt & Hemming* 2771 (K, holo. !, BR, EA, iso. !)

Main branches up to 2·5 m., erect or slightly curved. Thorns 0·4–2·5 cm., erecto-patent, mostly much shorter than the internodes. Inflorescences on short-shoots terminal and in the axils of thorns on the apical part of the branches. Male flowers including the 7–10 mm. long stipe 19–30 mm. ; hypanthium 6–12 mm. ; sepals 6–9 mm. ; petals 2·5–6 by 1·5–2·5 mm. ; filaments inserted 3·5–7 mm. above the base of the hypanthium, (2·5–)4–6 mm., connate up to 0·5 mm. ; anthers 4–6·5 mm. ; disk glands 0·7–1 mm. Female flowers including the 1–1·5 mm. long stipe 10–12 mm. ; hypanthium 3–4 mm. ; sepals 6–7 mm. ; petals 2–3 by 0·7–1 mm. ; staminodes inserted in the hypanthium 0·5–1 mm. above the base, 2–3·5 mm. ; disk glands 0·5 mm. ; pistil 7–8 mm. ; ovary 3·5–4·5 by 2·5–3 mm. ; ovules 5–9 per placenta. Fruit excluding the 1–2 mm. long gynophore 2–2·8 cm. Seeds 15–25 per capsule.

KENYA. Naivasha District : Mt. Suswa, 1 June 1963, *Glover* 3719 ! ; Kiambu District : Kikuyu to Narok via Kedong, 20 Jan. 1963, *Verdcourt* 3547 ! ; Masai District : Ngong Hills, Dec. 1938, *Bally* 8639 !
DISTR. **K**3, 4, 6 ; not known elsewhere
HAB. Deciduous and dry evergreen bushland ; 850–1700 m.

SYN. *A. pseudoglobosa* Verdc. subsp. *pseudoglobosa*, Bol. Soc. Brot., sér. 2, 38 : 98, t. 1–3, 7 (1964)

subsp. **curvata** (*Verdc.*) *de Wilde* in Blumea 17 : 180 (1969) & in Meded. Landbouwhogeschool Wageningen 71–18 : 133 (1971). Type : Tanzania, Mbulu District, Jungu–Aitjo, *Williams* 674 (EA, holo. !)

Main branches strongly curved or subscandent, up to 3 m. Thorns (0·5–)1–2 cm., patent or ± erecto-patent, ± as long as to much shorter than the internodes. Inflorescences on short-shoots terminal and in the axils of the thorns scattered along the apical ± 50 cm. of the branches. Male flowers including the 8–14 mm. long stipe 20–30(–35) mm. ; hypanthium 8(–10) mm. ; sepals 6–9 mm. ; petals 5–6 by 1·5–2·5 mm. ; filaments inserted in the hypanthium 2–3(–4) mm. above the base, 5–6 mm., free ; anthers 5·5–6 mm. ; disk glands 0·5–1 mm. Female flowers including the 1–1·5 mm. long stipe 10–12 mm. ; hypanthium 2–3 mm. ; sepals ± 7·5 mm. ; petals 2–2·5 by 0·7 mm. ; staminodes inserted at the base of the hypanthium, ± 2·5 mm. ; disk glands ± 0.2 mm. ; pistil 7–8 mm. ; ovary 3·5–4 by 2·5–3 mm. ; ovules 2–5 per placenta, 7–12 per ovary. Fruit excluding the 1–1·5 mm. long gynophore 2–2·8 cm. Seeds 7–12 per capsule.

TANZANIA. Masai District : Engaruka, foot of Mt. Loolmalasin, 9 July 1956, *Bally* 10669 ! ; Mbulu District : Jungu, 2 July 1956, *Bally* 10620 ! ; Kilimanjaro, 28 Apr. 1968, *Bigger* 1812 !
DISTR. **T**2 ; not known elsewhere
HAB. Deciduous and dry evergreen bushland, often in rocky places ; 900–1300 m.

SYN. *A. pseudoglobosa* Verdc. subsp. *curvata* Verdc. in Bol. Soc. Brot., sér 2, 38 : 101, t. 4, 5, 7 (1964)

NOTE. This subspecies is in several respects (e.g. habit, male flowers) intermediate between subsp. *globosa* and subsp. *pseudoglobosa*.

6. **A. venenata** *Forssk.*, Fl. Aegypt.-Arab.: 77 (1775) ; Engl. in E.J. 14 : 378, 374–375, 380 (1891) ; Harms in E. & P. Pf., ed. 2, 21 : 492, fig. 227 (1925) ; Blatter, Fl. Arab. 2 : 200 (1921) ; Dalz. in F.W.T.A., Append.: 50 (1937) ; F.P.S. 1 : 161, fig. 96 (1950) ; F.W.T.A., ed. 2, 1 : 202 (1954) ; de Wilde in Meded. Landbouwhogeschool Wageningen 71–18 : 133 (1971) & in U.K.W.F.: 165 (1974). Type : Yemen, *Forsskål* (C, holo. !)

Shrub or climber to 8 m., provided with a ± conical fleshy-woody main stem up to 2 m., glabrous. Leaf-blades grey to glaucous beneath, not punctate, shallowly to deeply 3–5(–7)-lobed, orbicular to ovate in outline, base cordate, top obtuse or rounded, 1·5–12 by 1·5–13 cm., 5-nerved from near base ; margin entire ; lobes semi-orbicular to lanceolate, rarely 3-lobed, obtuse, ± acute in juvenile specimens, up to 5 cm. ; petiole 1–8 cm. Gland at blade-base 1, median on the half-orbicular subpeltate base ; blade-glands

FIG. 6. *ADENIA VENENATA*—**1,** habit; **2,** leafy branchlet, × ⅔; **3,** female flowering branchlet, × ⅔; **4,** longitudinal section of female flower, × 3; **5,** part of hypanthium from same to show disk glands, staminodes and base of gynophore, × 6; **6,** male flowering branchlet, × ⅔; **7,** longitudinal section of male flower, × 2; **8,** part of hypanthium from same to show disk glands, bases of filaments and vestigial ovary, × 6; **9,** fruits, × ⅔; **10,** seed, × 4. 2–5, from *Bally & Smith* 14715; 6–8, from *Gillett* 13936; 9, 10, from *Gillett* 13390. Drawn by Mrs. M. E. Church.

0–6, submarginal. Stipules narrowly triangular, 0·5–1 mm., withering. Inflorescences mostly on short-shoots up to 5(–15) cm., rarely in the axils of normal leaves, peduncled up to 1.5 cm., (1–)3–5-flowered in ♂, 1(–3)-flowered in ♀; tendril 0(–1), up to 5 cm.; sterile tendrils simple, up to 12 cm. Male flowers tubiform-infundibuliform, including the (3·5–)15–33 mm. long stipe (18–)30–56 mm.; hypanthium tubiform, 9–15 by (1–)2–3 mm.; calyx-tube 0; sepals lanceolate, obtuse, (5–)6–9 mm., entire; petals oblong (to lanceolate), obtusish, 3–6 mm., 3–5-nerved, subentire; filaments 4–10 mm., free or up to 1 mm. connate, inserted 0·5–5 mm. above the base in the hypanthium; anthers 4–6 mm., obtuse; septa 0; corona 0; disk glands 0·5–1 mm., free or connate into a ring. Female flowers infundibuliform, including the 4–8 mm. long stipe 15–24 mm.; hypanthium 5–8 by 3(–4) mm.; calyx-tube 0; sepals oblong-lanceolate, obtuse, 5–8 mm., entire; petals oblong-lanceolate, 3–4 mm., 1–3-nerved, subentire; staminodes ± 3 mm., free, inserted at or near the base of the hypanthium; septa 0; corona 0; disk glands ± 0·5 mm., free; gynophore 2·5–5 mm.; ovary ovoid-oblong, 4–6 mm.; styles connate for 1–3 mm., free style-arms up to 0·5 mm.; stigmas subreniform, woolly-papillate, each 1·5–3 mm. in diameter. Fruits 1–2 per inflorescence, ± ovoid-ellipsoid, ± narrowed at the top, excluding the curved 7–12 mm. long gynophore 2–4·5 by 1·5–3 cm.; pericarp coriaceous. Seeds 15–35 per capsule, orbicular to broadly ovate, 4·5–6 mm. Fig. 6.

UGANDA. W. Nile District: Metuli, 25 Nov. 1941, *A. S. Thomas* 4056!; Bunyoro District: Bujenje, Feb. 1943, *Purseglove* 1290!; Teso District: Kyere, Mar. 1934, *Chandler* 1134!
KENYA. Northern Frontier Province: El Wak, 29 May 1952, *Gillett* 13390!; Baringo District: W. shore Lake Hannington, 20 Feb. 1971, *Mabberley* 723!; Lamu District: 48 km. N. of Lamu, Sept. 1956, *Rawlins* 109!
TANZANIA. Bukoba District: Nyashozi, Dec. 1931, *Haarer* 2460!; Mwanza District: Ukerewe I., *Conrads* 5205! & 5658!; Shinyanga, *Koritschoner* 1848!
DISTR. U1–4; K1–3, 5–7; T1; Yemen and N. tropical Africa from Nigeria to the Somali Republic
HAB. Wooded grassland, deciduous woodland and bushland; 0–1500 m.

SYN. *Modecca abyssinica* A. Rich., Tent. Fl. Abyss. 1: 297 (1847); Mast. in F.T.A. 2: 514 (1871). Type: Ethiopia, Tigre, Djeladjeranne, *Schimper* 242 (207/4) (P, holo.!, BM, FI, G, MPU, S, W, iso.!)

NOTE. The species is sometimes found planted near villages, because of its medicinal properties (see Dalziel, loc. cit.; de Wilde, loc. cit.)

7. **A. mannii** (*Mast.*) *Engl.* in E.J. 14: 375 (1891) & V.E. 3(2): 603 (1921); F.W.T.A., ed. 2, 1: 202 (1954); de Wilde in Meded. Landbouwhogeschool Wageningen 71–18: 144, fig. 21, 22 (1971). Type: Cameroun, Ambas Bay, *Mann* (K, holo.!)

Slender climber to 5 m., glabrous. Leaf-blades pale green, not punctate beneath, entire, elliptic to oblong or oblong-lanceolate, base acute to sub-cordate, top obtuse to acute, up to 1·5 cm. acuminate, 4–17 by 2–7(–8) cm., ± penninerved, sometimes with stronger nerves from near base, margin entire; petiole 0·5–2·5 cm. Glands at blade-base 2, on 2 ± hollowed auricles 1·5–3 mm. in diameter lateral at the top of the petiole; blade-glands scattered or submarginal, (0–)1–8 at either side. Stipules narrowly triangular, ± 0·5 mm., withering. Inflorescences peduncled for up to 12 cm., up to 30-flowered in ♂, up to 10-flowered in ♀; tendrils (0–)1 or 3, 0·5–15 cm.; sterile tendrils up to 20 cm. Male flowers tubiform, including the 0·5–3 mm. long stipe 15–23 mm.; hypanthium broadly cup-shaped, ± 5-saccate, 1·5–2·5 by 5–8 mm.; calyx-tube 8–12 mm.; calyx-lobes ± elongate-triangular, subobtuse to acute, subentire, 4–10(–12) mm.; petals linear to oblong, or spathulate or long-clawed, obtuse, 6·5–12 mm., (1–)3-nerved, fimbriate, inserted at the same level as the corona; filaments 4–7 mm., connate for up to 1·5 mm., inserted at the base of the hypanthium; anthers 5–6 mm., subobtuse, sometimes shortly apiculate; septa up to 1·5 mm. high; corona composed of fine threads 1–2 mm.; disk glands 1–1·5(–2) mm. Female flowers tubiform, including the 0–0·5 mm. long stipe 13–24 mm.; hypanthium broadly cup-shaped, ± 2 by (5–)7–9 mm.; calyx-tube 6–15 mm., calyx-lobes 5–7·5 mm.; petals linear, subacute, 4–10 mm., 1-nerved, entire, inserted at the same level as the corona; staminodes 3·5–7 mm.; septa up to 0·5 mm. high; corona threads fine, 1–2 mm.; disk glands 1–1·5(–2) mm.; gynophore ± 0·5 mm.; ovary ovoid-fusiform, 6–8 mm., ± 3-angular; styles connate for 0·2–2·5 mm., style arms 1·5–2·5 mm.; stigmas lacerate, woolly-papillate, each 3–4 mm. in diameter. Fruits 1–6 per inflorescence, broadly fusiform, up to 2 mm. beaked, (3- or)6-angular, excluding the ± 1 mm. long gynophore 2·5–3(–3·5) by 1·5–2 cm.; pericarp woody-coriaceous, 1–2 mm. thick. Seeds 20–30 per capsule, ovoid-oblong, 5–6(–7) mm.

UGANDA. Mengo District: Lake shore, 23 Jan. 1904, *Bagshawe* 124! & Entebbe, Kitu-
bulu Forest, Nov. 1922, *Maitland* 558!
DISTR. **U4**; W. and equatorial Africa, from Sierra Leone east to Uganda, south to
Zaire
HAB. Primary and secondary forest; 1050–1150 m.

SYN. *Modecca mannii* Mast. in F.T.A. 2: 516 (1871)
Adenia oblongifolia Harms in E. & P. Pf. III. 6A, Nachtr. 1: 255 (1897); F.W.T.A.,
ed. 2, 1: 202 (1954). Type: Cameroun, Johann Albrechts-Höhe, *Staudt* 621
(B, holo. †)
Modecca tenuispira Stapf in J.L.S. 37: 102 (1905). Type: Liberia, Sinoe Basin
(Greenville), *Whyte* (K, holo.!)
M. nigricans A. Chev., Expl. Bot. Afr. Occ. Fr. 1: 287 (1920), *nom. nud.*
Adenia tenuispira (Stapf) Engl., V.E. 3(2): 603 (1921), as " *tenuispica* "; F.W.T.A.,
ed. 2, 1: 202 (1954)

8. **A. panduriformis** *Engl.* in E.J. 14: 376 (1891); Harms in E. & P. Pf. III.6A: 84 (1893): Engl., V.E. 3(2): 604 (1921); de Wilde in Meded. Land-bouwogeschool Wageningen 71–18: 152 (1971). Type: Mozambique, Tete Province, *Kirk* in *Herb. Schweinfurth* (B, holo.†, K, iso.!)

Climber to 10 m., glabrous; stem ± terete, not winged or knobbly. Leaf-blades entire to 3(–7)-lobed, orbicular to ovate, base cordate or rarely truncate, top acute, up to 0·5 cm. acuminate, 4–13 by 3–12 cm., pinninerved or main nerves from near base, margin entire; lobes up to 4 cm.; petiole 0·5–5(–6) cm. Glands at blade-base 2, on 2 ± hollowed auricles 1·5–3 mm. in diameter lateral at the top of the petiole; no other glands present. Stipules oblong, acute, 0·5–1·5 mm. Inflorescences peduncled for up to 2 cm., up to 10-flowered in ♂, 1–5-flowered in ♀; tendril 0 or 1, 2–15 cm.; sterile tendrils up to 20 cm. Male flowers broadly tubiform-campanulate, including the 0·5–1·5 mm. long stipe 15–22 mm.; hypanthium broadly cup-shaped, ± 5-saccate, 3–5 by (8–)10(–14) mm.; calyx-tube 4–8 mm.; calyx-lobes elongate-triangular, subacute, 4–10 mm., finely undulate-crenate; petals ± broadly spathulate, top rounded to subacute, 3–5-nerved, dentate-fimbriate, including the 2–4 mm. long claw 7–8 by 3–4 mm., inserted at the same level as the corona; filaments 2·5–3 mm., connate for ± 1 mm., inserted at the base of the hypanthium; anthers 7·5–10 mm., obtuse, not apiculate; septa 0·5–1 mm. high; corona consisting of membranous appendages 0·5(–1) mm.; disk glands ± 1 mm. Female flowers broadly tubiform-campanulate, including the 0–1 mm. long stipe (10–)15–20 mm.; hypanthium 2–3 by (7–)10 mm.; calyx-tube 2–5 mm.; calyx-lobes narrowly triangular, acute, 7–14 mm., subentire; petals lanceolate to lanceolate-linear, acute, 5·5–7 mm., 1-nerved, crenulate-fimbriate towards the top; staminodes ± 3 mm.;

septa ± 0·5 mm. high; corona and disk glands as in ♂; gynophore 0·5–1 mm.; ovary subglobose, ± 5 by 4·5 mm.; styles connate for 1·5–2 mm., style-arms 1·5–2 mm.; stigmas subglobose, woolly-papillate, each ± 4 mm. in diameter. Fruits 1–3 per inflorescence, subglobose, excluding the 1–4(–5) mm. long gynophore 2·2–3·5(–4) by 2–3·5 cm.; pericarp woody-coriaceous, 1·5–2·5 mm. thick, green, turning yellow. Seeds 20–45 per capsule, ovoid-ellipsoid, ± 6 mm.

TANZANIA. Lindi District: 40 km. W. of Lindi, 20 Jan. 1935, *Schlieben* 5893 ! (see note)
DISTR. **T**8; Zambia, Rhodesia, Mozambique
HAB. Probably woodland or bushland; 250 m.

NOTE. The only collection from the Flora area, *Schlieben* 5893, is somewhat doubtful; the flowers resemble in a way those of certain forms of the very variable *A. rumicifolia*, but the anthers are blunt, not apiculate, a character distinctive for *A. panduriformis*. The locality of the specimen is rather far away from the main area of *A. panduriformis* which is in Zambia, Rhodesia and central Mozambique.

9. **A. rumicifolia** *Engl.*, V.E. 3(2): 603 (1921); Harms in N.B.G.B. 8: 296 (1923); de Wilde in Acta Bot. Neerl. 17: 292 (1968) & in C.F.A. 4: 221 (1970) & in Meded. Landbouwhogeschool Wageningen 71–18: 154 (1971) & in U.K.W.F.: 165 (1974). Type: Tanzania, Lushoto District, Gonja, *Engler* 3362 (B, holo.†, EA, iso. !)

Medium-sized to robust liana up to 20 (in W. and equatorial Africa up to 45) m., glabrous; older stems terete, 3–5-winged or with 5 rows of coarse fleshy tubercles, up to 10 cm. thick at base. Leaf-blades not punctate, entire or more rarely shallowly 3(–5)-lobed or coarsely sinuate, ovate to obovate, or orbicular to oblong, base acute to deeply cordate, or hastate, top acute to rounded, up to 2 cm. acuminate, 3·5–25 by 2·5–20 cm., pinninerved with 5–7 stronger nerves from near base; margin entire; petiole 1·5–15 cm. Glands at blade-base 2, in 2 ± hollowed auricles 1·5–6 mm. in diameter lateral at the top of the petiole; blade-glands submarginal, 0–6 on either half of the blade. Stipules triangular, 0·5–1 mm., withering. Inflorescences pedunceled for up to 12 cm., often sessile or subsessile on the lower part of the branches, up to 30-flowered in ♂, 1–6-flowered in ♀; tendril (0–)1, up to 20 cm., in sessile inflorescences tendril mostly absent; sterile tendrils simple or 3-fid, up to 25 cm. Male flowers broadly tubiform-campanulate, including the 0·5–2 mm. long stipe 16–37 mm.; hypanthium saucer-shaped, ± 5-saccate, 2·5–5 by 10–18 mm.; calyx-tube (5–)7–15 mm.; calyx-lobes ± elongate-triangular, subobtuse to acute, crenulate-laciniate, 7–14 mm.; petals mostly broadly spathulate, long-clawed, or elliptic-oblong, 4–13 by 2·5–5 mm., 3–5-nerved, lacerate-fimbriate, inserted at the same level as the corona; filaments 2–6 mm., connate 0·5–1·5 mm., inserted at the base of the hypanthium; anthers (6–)7–12 mm., acute, apiculate; septa 0·5–1·5 mm. high; corona consisting of woolly hairs 1–2 mm.; disk glands 2–3 mm. Female flowers broadly tubiform-campanulate, resembling ♂ flowers, including the ± 0·5 mm. long stipe 12–25(–30) mm.; hypanthium flattish, 2–3 by 7–15 mm.; calyx-tube (4–)5–10 mm.; calyx-lobes 7–16 mm.; petals linear to spathulate, subobtuse, (4–)5–9 mm., 1–3(–5)-nerved, laciniate-fimbriate, inserted at the same level as the corona; staminodes 3–6 mm.; septa, corona and disk glands as in ♂; gynophore (0·5–)1–2·5 mm.; ovary ovoid to ellipsoid, (3·5–)4–8 mm.; styles free or connate for up to 2 mm., style-arms 1·5–5 mm.; stigmas reniform, woolly-papillate, each 2·5–5 mm. in diameter. Fruits 1–4 per inflorescence, pear-shaped (obovate, tapering into the gynophore), obtuse, excluding the 1–5 mm. long gynophore 3–8 by 1·5–4·5 cm.; pericarp coriaceous outside, firmly fleshy inside, 4–10 mm. thick, smooth. Seeds 40–150 per capsule, ellipsoid to suborbicular, 3·5–5(–6) mm. Fig. 7.

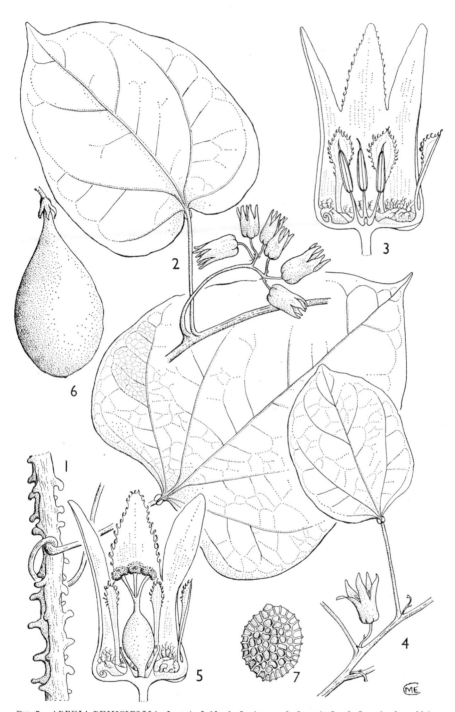

FIG. 7. *ADENIA RUMICIFOLIA*—**1,** part of older leafy stem, × ⅔; **2,** part of male flowering branchlet, × ⅔; **3,** longitudinal section of male flower, × 2; **4,** part of female flowering branchlet, × ⅔; **5,** longitudinal section of female flower, × 2; **6,** fruit, × 1; **7,** seed, × 4. 1, from *Faden* 69/2051; 2–5, from *Faulkner* 3446; 6, 7, from *Conrads* 5804. Drawn by Mrs. M. E. Church.

UGANDA. Ankole District: Ruizi R., Dec. 1950, *Jarrett* 131!; Mengo District: Kajansi Forest, Oct. 1937, *Chandler* 1997! & Mulange Hill, Nov. 1920, *Dummer* 4449!

KENYA. Embu District: Chuka, 26 Feb. 1922, *Fries* 1958a!; N. Kavirondo District: Kakamega Forest, 27 Nov. 1969, *Faden* 69/2051!; Kwale District: Makadara Forest, July 1939, *V. G. van Someren* 97!

TANZANIA. Buha District: Kasakela Reserve, 18 Nov. 1962, *Verdcourt* 3356!; Morogoro District: Chigurufumi Forest Reserve, Mar. 1955, *Semsei* 2038!; Rungwe District: Rutenganio [Rutangamio] Mission, Oct. 1910, *Stolz* 402!; Zanzibar I., Kidichi, 6 Dec. 1963, *Faulkner* 3324!

DISTR. U2, 4; K4, 5, 7; T1–4, 6, 7; Z; tropical Africa from Senegal east to S. Ethiopia, south to Angola and Mozambique

HAB. In moister places with a strong seasonal climate, in forest edges, semi-swamp forest, riverine forest and thicket; 0–1800 m.

SYN. *A. megalantha* Harms in P.O.A. A: 92 (1895), *nom. nud.*
    [*A. lobata* sensu Engl., P.O.A. B: 216 (1895); R. E. Fries, Wiss. Ergebn. Schwed. Rhod.-Kongo-Exped.: 157 (1914); A. & R. Fernandes in Garcia de Orta 6: 257, 659 (1958), pro parte; F.F.N.R.: 268 (1962), *non* (Jacq.) Engl.]
    A. lobata (Jacq.) Engl. var. *grandiflora* R. E. Fries, Wiss. Ergebn. Schwed. Rhod.-Kongo-Exped.: 157 (1914); Engl., V.E. 3(2): 604 (1921). Type: Zambia, Northern Province, Lake Bangweulu, *Fries* 1048 (UPS, holo.!)
    A. miegei Aké Assi in B.J.B.B. 31: 311, t. 1 (1961); de Wilde in Acta Bot. Neerl. 16: 233, fig. 1 (1967). Type: Ivory Coast, *Aké Assi* 5977 (P, holo.!, BR, iso.!)

NOTE. In my monograph (de Wilde, 1971) I have recognized two not strictly segregated varieties. Var. *miegei* (Aké Assi) de Wilde is more robust, with long-pyriform fruits (3·5–)5–8 cm., the attenuated part occupying at least one-third of the total length of the fruit, and occurs in rain-forests and clearings in west and central Africa. The type variety is less robust, and has short-pyriform fruits, 3–5 cm., the attenuated part occupying one-third or less of the total length of the fruit; this variety occurs predominantly in eastern equatorial Africa. In central Africa more or less intermediate forms are quite often found, and these are sometimes difficult to determine, especially when mature fruits are lacking. For this reason I have refrained from distinguishing varieties in the present treatment.

10. **A. schweinfurthii** *Engl.* in E.J. 14: 377 (1891); Harms in E. & P. Pf. III. 6A: 84 (1893) & ed. 2, 21: 491, fig. 218/G, H (1925) & in N.B.G.B. 8: 296 (1923); F.P.S. 1: 162 (1950); de Wilde in Acta Bot. Neerl. 17: 288, fig. 1 (1968) & in Meded. Landbouwhogeschool Wageningen 71–18: 159 (1971) & in U.K.W.F.: 165 (1974). Type: Zaire, Ubangi-Uele, Munza, *Schweinfurth* 3485 (K, lecto.!, P, isolecto.!)

Climber up to 20 m., glabrous; older stem terete or 2–4(–5)-angular or -winged, rarely with small tubercles. Leaf-blades pale green beneath, not punctate, entire or rarely shallowly 3-lobed, ovate or obovate to orbicular, base acute to cordate, top acute, up to 1·5 cm. acuminate, 5–12(–15) by (3·5–)4–12 cm., pinninerved with 5–7 stronger nerves from near base; margin entire; petiole 2–6(–11) cm. Glands at blade-base 2, in 2 hollowed auricles 2·5–5 mm. in diameter lateral at the top of the petiole; blade-glands sub-marginal, 0–1(–2) at either side of the blade. Stipules subtriangular, ± 0·5 mm., withering. Inflorescences peduncled up to 6 cm. or sessile on the basal part of the branches, 2–20-flowered in ♂, 1–4(–10)-flowered in ♀; tendril (0–)1, up to 10 cm.; sterile tendrils up to 20 cm. Male flowers broadly tubular-campanulate, including the 0·5–1·5 mm. long stipe 8–15(–20) mm.; hypanthium broadly crateriform, ± 5-saccate, 2–3(–4) by 4–8(–10) mm.; calyx-tube 2–4(–6) mm.; calyx-lobes elongate-triangular, subacute, 4–8(–12) mm., crenulate-laciniate; petals spathulate, obtuse, 5–9 by 3–4 mm., 3(–5)-nerved, laciniate-fimbriate, inserted at the same level as the corona; filaments 3·5–7 mm., connate for 1–2 mm., inserted at the base of the hypanthium; anthers 3·5–5(–6) mm., subobtuse, up to 1 mm. apiculate; septa 1–2 mm. high; corona threads fine, 0·5–1 mm.; disk glands ± 1 mm. Female flowers broadly tubular-campanulate, including the 0–0·5 mm. long stipe 10–14 mm.; hypanthium ± 2 by 6–8 mm.; calyx-tube 3–4 mm.;

calyx-lobes 5–8 mm.; petals lanceolate to narrowly spathulate, obtuse, ± 6 mm., (1–)3-nerved, fimbriate-lacerate, inserted at the same level as the corona; staminodes 3·5–4 mm.; septa 0·5–1 mm. high; corona threads ± 1 mm.; disk glands 1–2 mm.; gynophore ± 1(–1·5) mm.; ovary subglobose to broadly ovoid, 4·5–5 mm.; styles ± 2 mm., free; stigmas subreniform, woolly papillate, each ± 3 mm. in diameter. Fruits 1–2 per inflorescence, subglobose to ellipsoid, excluding the 5–10 mm. long gynophore 3·5–5(–6) by 2·5–4 cm.; pericarp thickly coriaceous with woody endocarp, fleshy, (4–)5–10 mm. thick. Seeds 45–100 per capsule, orbicular to ovoid, ± 4–4·5 mm.

UGANDA. W. Nile District: Logiri, Mar. 1935, *Eggeling* 1889!; Bunyoro District: Hoima, 27 Feb. 1907, *Bagshawe* 1514! & Victoria Falls, Foweira, 10 Apr. 1907, *Bagshawe* 1571!
KENYA. Nandi District: Kaimosi, SW. of Yala R. bridge, 18 Apr. 1965, *Gillett* 16701!
TANZANIA. Bukoba District: Kikuru Forest, Sept.–Oct. 1935, *Gillman* 443!; Mwanza District: Ukerewe I., *Conrads* 5339! & 6297!; Mbulu District: Lake Manyara National Park, Mto wa Ukindu, 22 Nov. 1963, *Greenway & Kanuri* 11067!
DISTR. **U**1, 2; **K**3; **T**1, 2; Cameroun, Central African Republic, Zaire, Sudan
HAB. Primary and secondary forest, riverine forest, semi-swamp forest; 950–1700 m.

SYN. *Modecca koutiensis* A. Chev., Études Fl. Afr. Centr. Fr. 1: 135 (1913). Types: Central African Republic, *Chevalier* 7740, 8314 & 8314 bis (all P, syn.!)
　　　Adenia koutiensis (A. Chev.) Obaton in Ann. Sci. Nat., sér. 12, 1: 142 (1960), comb. non rite publ.

11. **A. stenodactyla** *Harms* in N.B.G.B. 8: 297 (1923); de Wilde in Meded. Landbouwhogeschool Wageningen 71–18: 163 (1971). Type: Tanzania, Songea District, Matagoro Hills, *Busse* 826b (B, holo.†)

Climber to ± 2·5 m., from a tuberous rootstock, glabrous. Leaves greyish green, sometimes spotted beneath, very deeply 5-parted or 5-foliolate, suborbicular in outline, (2–)4–10(–25) by 4–20(–25) cm., 5-nerved from base; leaflets oblong to linear, base long-acute, top subobtuse to acute, up to 1·5 cm. acuminate, entire to deeply 3–8(–16)-lobed, 2–16 cm.; margin entire; petiolules up to 1·5 cm.; petiole 0·5–5 cm. Glands at blade-base 2(–4), on 2 separate or contiguous wart-like appendages at the top of the petiole; blade-glands (0–)2–4(–16), submarginal or partly scattered. Stipules linear, ± 1 mm. Inflorescences peduncled for 1–8 cm., up to 10-flowered in ♂, 1–3-flowered in ♀; tendril (0–)1, 2–6 cm.; sterile tendrils up to 12 cm. Flowers often reddish veined or striped. Male flowers broadly urceolate, including the 4–8 mm. long stipe 25–40 mm.; hypanthium broadly cup-shaped, ± 5-saccate, 4–7 by 8–13 mm.; calyx-tube 12–20 mm.; calyx-lobes ovate-triangular, subobtuse, 4–8 mm., lacerate-denticulate; petals oblong-spathulate, obtuse, ± clawed, 6–10 mm., 5-nerved, lacerate-dentate, inserted at the same level as or up to 2 mm. above the corona; filaments 4–7·5 mm., connate for 1–3 mm., inserted at the base of the hypanthium; anthers 6–8(–10) mm., obtuse or subacute; septa 1–3 mm. high; corona threads fine, 0·5–1·5 mm.; disk glands 3–3·5 mm. Female flowers tubiform-campanulate, including the 2–3(–4) mm. long stipe 15–18 mm.; hypanthium broadly cup-shaped, 2–2·5 by 5–8 mm.; calyx-tube 6–8 mm.; calyx-lobes ovate-triangular, subacute, entire, 5–7 mm.; petals oblong-spathulate, obtuse, ± 6 mm., 5-nerved, denticulate-lacerate, inserted at the same level as the corona; staminodes ± 4 mm.; septa ± 1 mm. high; corona threads sparse, 1–2 mm.; disk glands ± 1 mm.; gynophore ± 2 mm.; ovary ellipsoid, ± 4 mm.; styles connate for 2–2·5 mm.; stigmas sessile, woolly-papillate, each ± 2 mm. in diameter. Hermaphrodite flowers resembling ♂ flowers. Fruits 1–2 per inflorescence, ellipsoid, excluding the 8–15 mm. long gynophore (2·5–)3–6 by 2–3·5 cm.; pericarp coriaceous, red. Seeds 40–60 per capsule, obovate, 5–6·5 mm. Fig. 5/3, p. 21.

TANZANIA. Ufipa District: near Lake Sundu, 10 Dec. 1958, *Richards* 10295!; Dodoma District: Kazikazi, 23 Dec. 1932, *B. D. Burtt* 3824!; Songea District: 19 km. W. of Songea, 30 Dec. 1955, *Milne-Redhead & Taylor* 7789! & 7790!
DISTR. **T**4, 5, 7, 8; Zambia
HAB. Deciduous woodland, bushland and thicket, grassland, rocky hillsides; 900–2100 m.

SYN. *A. angustisecta* Engl., V.E. 3 (2): 605 (late 1921), *non* Burtt Davy (Sept. 1921), *nom. illegit.* Type: Tanzania, Songea District, Matagoro Hills, *Busse* 826b (B, holo. †)
A. stenodactyla Harms var. *kondensis* Harms in N.B.G.B. 8: 298 (1923) & 13: 426 (1938). Type: Tanzania, Mbeya/Chunya Districts, Usafwa, *Stolz* 2355 (B, holo.†, BM, BR, K, P, Z, iso.!)

12. **A. dolichosiphon** *Harms* in N.B.G.B. 13: 425 (1936); de Wilde in Meded. Landbouwhogeschool Wageningen 71–18: 165 (1971). Type: Tanzania, Lindi District, Nambilanje, *Schlieben* 6001 (B, holo!, BR, K, LISC, iso.!)

Herbaceous climber to 5 m., glabrous, growing from a tuber. Leaf-blades pale greenish, not punctate beneath, entire, broadly ovate to triangular, rarely lanceolate, base cordate to truncate or hastate, top obtuse to acute, 2–10 by 3–8 cm., with 3–5(–7) main nerves from near base or ± pinninerved, margin entire; petiole 0·5–5·5 cm. Glands at blade-base 2, contiguous, situated on the subspathulate or ± 2-lobed short-peltate blade-base; no other glands. Stipules narrowly triangular, 1–2 mm. Inflorescences sessile or peduncled up to 5 cm., 1–15-flowered in ♂, 1–3-flowered in ♀; tendril 1, simple, 3–12 cm.; sterile tendrils up to 15 cm. Male flowers tubiform, including the 4–9 mm. long stipe 40–75 mm.; hypanthium ± cup-shaped, tapering, 1·5–3 mm.; calyx-tube 30–55 by 5–9 mm.; calyx-lobes ovate-oblong, subobtuse, 5–7·5 mm., crenulate or fimbriate; petals linear, acute, 30–45 mm., 1–3-nerved, densely ± 2 mm. fimbriate, inserted at the same level as or up to 2 mm. above the corona; filaments 7–10 mm., connate for 0·5–1 mm., inserted at the base of the hypanthium or on a short androgynophore; anthers 7–8·5 mm., obtuse; septa ± 2 mm. high; corona threads fine, 1–1·5 mm.; disk glands ± 2 mm. Female flowers not known. Fruits 1–2 per inflorescence, subglobose, excluding the 1–2 mm. long gynophore 4–5 by 4 cm.; pericarp thickly coriaceous. Seeds 15–20 per capsule, ovate, ± 7 mm.

TANZANIA. Rufiji R., 4 Feb. 1931, *Musk* 42!; Lindi District: Tendaguru, *Migeod* 723! & Nambilanje, 16 Feb. 1936, *Schlieben* 6001!
DISTR. **T**6, 8; Mozambique
HAB. Deciduous woodland and bushland; 50–200 m.

NOTE. In the original description by Harms the stem is described as minutely puberulous to subglabrous; however, all plants seen, including the type specimens, are entirely glabrous.
Half-grown hermaphrodite flowers have been seen in *Musk* 42.

13. **A. metriosiphon** *de Wilde* in Blumea 17: 179 (1969) & in Meded. Landbouwhogeschool Wageningen 71–18: 166 (1971) & in U.K.W.F. 165 (1974). Type: Kenya, Nairobi, Kirichwa Ndogo valley, *Bally* 11956 (K, holo.!, EA, iso.!)

Climber to 6 m., glabrous, growing from a tuberous rootstock. Leaf-blades grey-green, often reddish nerved, not punctate beneath, entire, ovate to suborbicular, base cordate to subtruncate, top obtuse to subacute, 2–9(–17) by 1·5–6(–11) cm., 5(–7)-nerved from base to ± pinninerved; margin entire; petiole 1–6(–9) cm. Glands at blade-base 2, separate or ± contiguous, on 2 small auricles which are ± connate over the top of the petiole; no other

glands present. Stipules triangular to lanceolate, acute, 0·5–2·5 mm.
Inflorescences sessile or peduncled for up to 5 cm., up to 5-flowered in ♂,
1–2-flowered in ♀; tendril 0 or 1, (0·5–)3–8 cm.; sterile tendrils up to 15 cm.
Male flowers tubiform-urceolate, including the 4–4·5 mm. long stipe
20–35(–38) mm.; hypanthium cup-shaped, tapering, 0·5–1 by 3 mm.;
calyx-tube 10–25 by 5–11 mm., lobes ovate to triangular, subacute, 4–8 mm.,
crenulate-laciniate; petals linear, subacute, 10–18 mm., 3-nerved, densely
feather-like, 1–2 mm. fimbriate, inserted 1–5 mm. above the corona; filaments 2·5–3 mm., free, on a short androgynophore; anthers 5·5–7 mm.,
obtuse; septa 0; corona threads 0·5–1 mm.; disk glands 1–1·5 mm. Female
flowers shortly tubiform, including the 0·5–1·5 mm. long stipe 15–20 mm.;
hypanthium 1–1·5 by 4 mm.; calyx-tube 10–14 by 8–10 mm., lobes triangular, subacute, 3–4 mm., crenulate-fimbriate; petals linear, acute,
(7–)8–10 mm., 1-nerved, fimbriate or not, inserted ± 2 mm. above the
corona; staminodes ± 2 mm.; septa 0; corona threads 0·5–1 mm.; disk
glands 0·5–1 mm.; gynophore 1–1·5 mm.; ovary subglobose, 4–5·5 mm.;
styles connate for 1 mm., style-arms 1–1·5 mm.; stigmas much branched,
± woolly-papillate, each 2·5–3·5 mm. in diameter. Fruits 1 per inflorescence,
subglobose, excluding the 2–5 mm. long gynophore 4–5 by 3·5–4·5 cm.;
pericarp thick, coriaceous outside, spongy inside. Seeds ± 40 per capsule,
ovoid-ellipsoid, ± 6–9 mm.

KENYA. Kiambu, 5 Nov. 1932, *Napier* 3558!; Machakos District: Athi R., 32 km.
beyond Thika, 31 Oct. 1938, *Bally* 8454!; Masai District: Garabani Hill, 10 Mar.
1940, *V. G. van Someren* 64!
DISTR. **K4, 6**; not known elsewhere
HAB. Dry evergreen forest and bushland, streamsides; 1350–2100 m.

14. **A. lindiensis** *Harms* in N.B.G.B. 13: 425 (1936); de Wilde in Meded.
Landbouwhogeschool Wageningen 71–18: 173 (1971). Type: Tanzania,
Lindi District, Rondo [Muera] Plateau, *Schlieben* 6066 (B, holo.†, LISC,
iso.!)

Climber to 5 m., glabrous. Leaf blades pale green, not punctate beneath,
entire to deeply (2–)3(–5)-lobed, oblong-lanceolate to broadly ovate or
suborbicular in outline, base rounded to cordate or hastate, top acute, up to
1·5 cm. acuminate, 4–17 by 2·5–16 cm., ± pinninerved or 3–5-nerved from
near base; lobes triangular to oblong, 1–12 cm.; margin entire or irregularly
sinuate; petiole 1·5–9 cm. Glands at blade base 2, separate or contiguous,
either on 2 small separate auricles lateral at the top of the petiole, or on the
2–6 mm. wide peltate blade-base; blade glands 2–20, submarginal or ±
scattered; marginal glands minute, 0–4. Stipules narrowly triangular,
± 1–1·5 mm., withering. Inflorescences peduncled for (0·5–)1–5 cm.,
1–5(–10)-flowered in ♂, ♀̣ and ♀; tendril (0–)1, 2–6 cm.; sterile tendrils up to
10 cm. Male flowers infundibuliform, including the ± 5 mm. long tapering
stipe ± 22 mm.; hypanthium cup-shaped, ± 2·5 by 2·5–3 mm.; calyx-tube
9–10 by ± 5 mm., lobes triangular-oblong, 5–6 mm.; petals obovate-oblong,
± 4·5–5 mm., 3-nerved, serrate, not long-fimbriate, inserted about half-way
in the calyx-tube ± 4·5 mm. above the hypanthium; filaments ± 7 mm.,
connate for ± 1·5 mm.; anthers 3–3·5 mm., obtuse, inserted at the base of
the hypanthium; septa ± 1·5 mm. high; corona ± 0; disk glands ± 1·5 mm.
Hermaphrodite flowers infundibuliform, including the ± 10 mm. long tapering stipe ± 30 mm.; hypanthium including calyx-tube 12–13 mm., ±
8 mm. wide at throat; calyx-lobes ovate-oblong, obtuse, entire, ± 8 mm.;
petals linear, 1–1·5 mm., 1-nerved, entire, inserted 7–8 mm. above the base
of the hypanthium; filaments 7–8 mm., connate for ± 0·3 mm., inserted at
the base of the hypanthium; anthers 1·5–2 mm.; septa ± 0·3 mm.; corona

0; disk glands ± 2 mm.; pistil see ♀ flowers. Female flowers ± infundibuliform, including the 1·5–2·5 mm. long stipe 10–15 mm.; hypanthium cup-
shaped, 0·5–1·5 by 2 mm.; calyx-tube 4–5 by 2–5 mm., lobes triangular to
oblong, subacute, 3–6 mm., entire; petals linear, 1·5–3 mm., 1-nerved,
± serrulate, inserted 0·5–3 mm. above the base of the hypanthium; staminodes ± 2 mm.; septa up to 1 mm. high; corona 0 or consisting of minute
hairs ± 0·1 mm.; disk glands 0; gynophore 2–5 mm.; ovary subglobose to
ellipsoid-oblong, 3–7 mm.; styles connate for 0·5–1 mm., style-arms ±
0·7–1·5 mm.; stigmas .subreniform, laciniate, woolly-papillate, each ±
2·5 mm. in diameter. Fruits 1 per inflorescence, ellipsoid to subglobose,
excluding the 8–10 mm. long gynophore 3–5 by 2·5–5 cm.; pericarp coriaceous,
spongy inside. Seeds ± 40 per capsule, subglobose, ± 5 mm.

KENYA. Kwale District: Shimba Hills, Makadara Forest, July 1939, *V. G. van Someren*
87!
TANZANIA. Tanga District: Ngomeni, 31 July 1953, *Drummond & Hemsley* 3562! &
Kange Limestone Gorge, 13 Nov. 1955, *Milne-Redhead & Taylor* 7288!; Morogoro
District: Uluguru Mts., 30 Dec. 1970, *Wingfield* 2165! & Lusunguru Forest, 31 Mar.
1953, *Drummond & Hemsley* 1930!
DISTR. **K7**; **T3, 6, 8**; not known elsewhere
HAB. Shrub-layer and edges of evergreen forest and associated bushland; 0–1200 m.

SYN. *A. lindiensis* Harms var. *submarginalis* de Wilde in Meded. Landbouwhogeschool
Wageningen 71–18: 175 (1971). Type: Tanzania, Lushoto District, Sangerawe–
Kwamkoro, *Zimmermann* in *Herb. Amani* 6577 (EA, holo.!, BM, K, iso.!)

NOTE. In my monograph of *Adenia* (de Wilde, loc. cit.), two varieties not maintained
in the present treatment, were distinguished on rather weak vegetative characters
mainly concerning the position of the glands on the blade-base; these either situated
on two separate auricles in var. *lindiensis*, or situated on the peltate blade base in
var. *submarginalis*. *Drummond & Hemsley* 1930 is a specimen with the leaves in all
characters intermediate between the varieties, but the glands situated on separate
auricles formerly referred it to var. *lindiensis*. It is known from certain other species,
e.g. *A. volkensii*, with two separate basal blade glands in the adult stage, that leaves
in the juvenile stage have distinct peltate blade-bases.
    Certain specimens of *A. lindiensis* resemble *A. schliebenii*, a species differing by the
slightly hairy leaves, the dentate leaf margin, and by some flower characters.
    Schlieben 6066, from the Rondo Plateau, has hermaphrodite flowers.

15. **A. schliebenii** *Harms* in N.B.G.B. 13: 426 (1936); A. & R. Fernandes in
Garcia de Orta 6: 257, fig. 13 (1958); de Wilde in Meded. Landbouwhoge-
school Wageningen 71–18: 175 (1971). Type: Tanzania, Rondo [Muera]
Plateau, *Schlieben* 5975 (B, holo.!, BM, BR, HBG, LISC, Z, iso.!)

Climber up to 5 m., growing from a tuberous rootstock, glabrescent.
Leaf-blades pale green, densely reddish punctate, finely pubescent especially
on the nerves beneath, entire to deeply 3–5-lobed, suborbicular in outline,
base cordate to subtruncate, top acute, up to 1 cm. acuminate, 5–12 by
4–13 cm., ± 5-nerved from base; lobes elliptic to oblong, ± constricted at
base, 1–8 cm.; margin remotely up to 3 mm. dentate; petiole 2–6·5 cm.
Glands at blade-base 2, on 2 small auricles lateral at the top of the petiole;
no other glands present. Stipules triangular, acute-acuminate, ± 1 mm.
Inflorescences peduncled for 0·5–6 cm., 3–15-flowered in ♂; tendril 1·5–5
cm.; sterile tendrils up to 15 cm. Male flowers tubular-infundibuliform,
including the 3–6 mm. long stipe 25–35 mm.; hypanthium, including calyx-
tube, (15–)18–24 by (6–)8–10 mm., lobes ovate-triangular, subacute, 4–5·5
mm., serrulate; petals lanceolate, 3–5 mm., 3-nerved, densely fimbriate,
inserted 11–15 mm. above the base of the hypanthium; filaments 10–21 mm.,
connate for 1–2 mm., inserted at the base of the hypanthium; anthers
(3–)4–7 mm., obtuse; septa 0; corona 0; disk glands ± 1·5 mm. Female
flowers and fruit not known.

TANZANIA. Lindi District: Rondo [Muera] Plateau, 14 Feb. 1935, *Schlieben* 5975!
DISTR. **T8**; Mozambique
HAB. Bushland; ± 500 m.

NOTE. This is one of the few pubescent species of *Adenia*. It is also characterized by
the finely red-brown spotted leaves. The leaves are reminiscent of those of *A. volkensii*
(which are also pubescent), and of *A. lindiensis*, in which they are glabrous with an
entire (not dentate) margin. Fresh flowers are yellowish, reddish brown punctate.

16. **A. ellenbeckii** *Engl.*, V.E. 3(2): 606, fig. 270 (1921); Harms in N.B.G.B.
8: 298 (1923) & in E. & P. Pf., ed. 2, 21: 492, fig. 225 (1925); de Wilde in
Meded. Landbouwhogeschool Wageningen 71–18: 178 (1971) & in Acta
Bot. Neerl. 21: 564, fig. 2/a–d, 3/f (1972) & in U.K.W.F.: 166 (1974). Types:
Ethiopia, Borana, Wai-Wai, *Ellenbeck* 2133 & Somali Republic (S.), Juba,
Umfuda, *Ellenbeck* 2291 (both B, syn.†)

Shrub or herb up to 1·5 m., shoots climbing or suberect on a woody
succulent stem up to 60 cm. arising from a tuberous rootstock, pubescent,
rarely subglabrous. Leaf-blades pale green, not punctate beneath, mostly
pubescent especially on the nerves, entire to mostly deeply palmately
3–5(–7)-lobed or pinnatifid, elliptic or ovate to suborbicular in outline, base
acute to cordate, 2–17 by 1·5–11 cm., with 3–5 main nerves from near base
or ± pininnerved; lobes ovate or obovate to oblong, 1–8 cm., top obtuse to
acute; margin subentire or variously dentate or dissected, the small teeth
ending in a blackish mucro; petiole (0·5–)1–7 cm. Glands at blade-base
2, sessile or sometimes on 2 small auricles on each side at the very top of the
petiole; blade glands 0–6, submarginal. Stipules triangular, dark brown
dotted, 1–2·5 mm. Inflorescences sessile, 1–10-flowered in ♂, 1–3-flowered
in ♀; tendril 0–1, 2–10 cm.; sterile tendrils up to 10 cm. Flowers subglabrous. Male flowers tubiform, including the 3–4 mm. long stipe 20–50 mm.;
hypanthium including calyx-tube 14–45 by 3–6·5 mm.; calyx-lobes ovate-
oblong, obtuse, 3–6 mm., woolly-fimbriate; petals lanceolate-linear, acute,
5–8(–11) mm., 1(–3)-nerved, woolly-fimbriate, inserted in the calyx-tube
3–7 mm. below the throat; filaments 6–12 mm., free or connate up to 3 mm.,
inserted at the base of the hypanthium or on a short androgynophore;
anthers 5–8 mm., obtuse; septa 0; corona 0 or consisting of but a few hairs;
disk glands ± 2·5 mm., rarely absent. Female flowers tubiform-urceolate,
including the 1–3 mm. long stipe 12–30(–35) mm.; hypanthium including
calyx-tube 8–25 by 4·5–8(–9) mm.; calyx-lobes ovate-oblong, subobtuse,
3–6 mm., woolly-fimbriate; petals linear, 4–9 mm., 1-nerved, fimbriate or
not, inserted in the calyx-tube 1–5 mm. below the throat; staminodes
2–7·5 mm.; septa 0; corona 0; disk glands ± 1 mm., free or rarely connate
into a ring; gynophore 1–2·5 mm.; ovary subglobose to ellipsoid, 3–7·5 mm.;
styles connate for 0–4 mm., free style arms 0·5–4 mm.; stigmas much
divided, hairy-papillate, each 2–5 mm. in diameter. Fruits 1 per inflorescence, subglobose, excluding the 1–2 mm. long gynophore 2–4·5(–5) by
2–4(–4·5) cm.; pericarp coriaceous, ± spongy or fleshy inside. Seeds 10–20
per capsule, ovate, 6·5–8 mm.

UGANDA. Karamoja District.: Kanamugit, *Eggeling* 2999! & 40 km. N. of Kacheliba
to Karamoja, 8 May 1953, *Padwa* 77! & Lokapeliethe, 29 Oct. 1939, *A.S. Thomas*
3104!
KENYA. Northern Frontier Province: Furroli Mt., 17 Sept. 1952, *Gillett* 13926!;
Machakos District: Mtito Andei, 2 Nov. 1966, *Greenway & Duvigneaud* 12622!;
Kitui District: Ngomeni, 31 Nov. 1945, *D.C. Edwards* 32!
TANZANIA. Mwanza District: Nyambiti, 17 Mar. 1953, *Tanner* 1292! & Issansu,
Kohl-Larsen 10!
DISTR. **U**1; **K**1, 2, 4, 6; **T**1, ?2; Ethiopia, Somali Republic
HAB. Deciduous bushland, often in rocky places, on laterite and clay soils; 150–1200
(?–1500) m.

SYN. *A. toxicaria* Harms in N.B.G.B. 13: 426 (1936). Type: Tanzania, Issansu, *Kohl-Larsen* 10 (B, holo.†)
    A. vitifolia Hutch. & Bruce in K.B.: 98 (1941); Cufod. in B.J.B.B. 29, Suppl.:
    600 (1959). Type: Somali Republic (N.), boundary, *Gillett* 4202 (K, holo.!,
    FI, P, UPS, iso.!)

NOTE. Related to *A. volkensii*, but differing by the presence of tendrils, by the inflorescences which are always sessile, and by the narrow flowers in which the corona is lacking or almost lacking.

    The specimen *Bally* 5488 is monoecious; it contains fruit remnants near the base of the plant, whereas male flowers are present in the upper part.

    Juvenile leaf-forms sometimes have a peltate blade-base. Several times reported as poisonous. The juice of the fruit in meat has been used to poison hyenas.

17. **A. keramanthus** *Harms* in E. & P. Pf. III. 6A: 84 (1893) & ed. 2, 21:
492 (1925) & in P.O.A. C: 281 (1895) & in E.J. 24: 177 (1897); Engl., V.E.
3 (2): 595, 605, fig. 269 (1921); de Wilde in Meded. Landbouwhogeschool
Wageningen 71-18: 180 (1971) & in U.K.W.F.: 165 (1974). Type: Zanzibar,
Kirk (K, holo.!)

Sparingly branched herb or shrub with succulent stems to ± 1 m., arising
from a tuberous rootstock; plant densely rusty-brown hairy; no tendrils.
Leaf-blades some pale green, sometimes punctate near the margin beneath,
not lobed, ovate to orbicular, base cordate to truncate, top subacute to
rounded, 1·5–15 by 1–14 cm., 5-nerved from base to ± pinninerved; margin
subentire or up to 5 mm. dentate; petiole 1–12 cm. Glands at blade-base
2, on the up to 10 mm. wide peltate blade-base; blade-glands 0; marginal
glands minute, as blackish mucros on the teeth. Stipules linear, sometimes
lacerate, 3–10 mm., pubescent. Inflorescences sessile, up to 10-flowered in
♂, 1–2-flowered in ♀; tendrils 0. Male flowers urceolate, including the
1–3·5 mm. long stipe 17–26 mm.; hypanthium broadly cup-shaped, 1·5–2 by
8–10 mm.; calyx-tube 8–17 mm., lobes ovate or ovate-oblong, obtusish,
3–5 mm., woolly-fimbriate; petals lanceolate to linear, 5–7 mm., 1-nerved,
fimbriate, inserted 2–9 mm. above the corona; filaments 3–5 mm., up to
1·5 mm. connate, inserted at the base of the hypanthium; anthers 6–7 mm.,
obtuse; septa 0; corona threads sparse, 1–2 mm., rarely absent; disk glands
1–2 mm. Female flowers urceolate, resembling ♂ flowers, including the ±
1·5 mm. long stipe 18–22 mm.; hypanthium 1–1·5 by 8–12 mm.; calyx-tube
12–14 mm., lobes 4–5 mm.; petals linear, ± 3 mm., 1-nerved, fimbriate,
inserted 6–7 mm. above the corona; staminodes ± 3·5 mm.; septa 0; corona
consisting of a few hairs ± 1 mm.; disk glands ± 1 mm.; gynophore 1–1·5
mm.; ovary subglobose to ellipsoid, 5–6 mm.; styles connate for ± 3 mm.,
style-arms 1–1·5 mm.; stigmas ± reniform, laciniate-papillate, each 4–5 mm.
in diameter. Fruits 1 per inflorescence, subglobose, excluding the 1–2 mm.
long gynophore 3·5–5 by 3–4·5 cm.; pericarp coriaceous, sometimes spongy
inside, crimson or purplish when fresh. Seeds 20–40 per capsule, broadly
ovoid to ellipsoid, 7(–9) mm.

KENYA. Machakos District: Kibwezi, 4 July 1969, *Faden & Evans* 69/806!; Teita
    District: Voi–Moshi road, 9 Aug. 1955, *Ossent* 98!; Kwale District: Mackinnon Road,
    9 Sept. 1953, *Drummond & Hemsley* 4233!
TANZANIA. Kilimanjaro, 20 Nov. 1966, *Gilbert* 4883!; Lushoto District: Mombo, 20
    Aug. 1950, *Verdcourt & Greenway* 330!; Handeni District: Sindeni [Zindeni], 12 Oct.
    1933, *B.D. Burtt* 4887!; Uzaramo District: 96 km. W. of Dar es Salaam, 17 Aug. 1969,
    Harris 3132!; Zanzibar I., Feb. 1874, *Hildebrandt* 1198!
DISTR. **K**4, 7; **T**2, 3, 6; **Z**; not known elsewhere
HAB. Deciduous woodland and bushland, dry evergreen and coastal bushland; 0–1000 m.

SYN. *Keramanthus kirkii* Hook. f. in Bot. Mag., t. 6271 (1876). Type: as for species

NOTE. The flowers are often produced before the leaves are fully developed. Leaves
    and roots are reported as a snake-bite treatment.

18. **A. volkensii** *Harms* in P.O.A. C: 281 (1895); Engl., V.E. 3(2): 606 (1921); Harms in E. & P. Pf., ed. 2, 21: 492 (1925); Verdc. & Trump, Common Poisonous Pl. E. Afr.: 37, fig. 3 (1969); de Wilde in Meded. Landbouwhogeschool Wageningen 71–18: 182, fig. 29 (1971) & in U.K.W.F.: 165 (1974). Type: Tanzania, Kilimanjaro, Kahe, *Volkens* 2174 (B, holo.†, BM, iso.!)

Somewhat woody shrub or herb up to 1·5 m., the erect annual shoots 20–60 cm., arising from a tuberous rootstock or a ± succulent stem; no tendrils; plant, especially the veins of the leaves, pubescent, rarely glabrous. Leaf-blades grey-green, sometimes punctate, sometimes reddish veined beneath, subentire to deeply 3–7-lobed or -dissected, base cordate to truncate, 3–16 by 3–14 cm., 3–5(–7)-nerved from near base to ± pinninerved; lobes oblong to lanceolate, acute, 2–12 cm.; margin finely dentate or up to 2 cm. dissected; petiole 1·5–10 cm. Glands at blade-base 2, situated on 2 small auricles at the top of the petiole; blade glands 2–6, submarginal; marginal glands as blackish mucros on the teeth. Stipules triangular, acute, gland-dotted, ± 1 mm. Inflorescences peduncled for up to 0·5 cm., 1–6-flowered in ♂, 1–2(–3)-flowered in ♀, without tendrils. Flowers glabrous. Male flowers urceolate, including the 2–8 mm. long stipe 20–35(–45) mm.; hypanthium broadly cup-shaped, ± 5-saccate, 3–6 by 10–18(–20) mm.; calyx-tube 12–20 mm., 7–12 mm. wide at throat; calyx-lobes ovate-triangular, obtusish, (3–)4–7(–9) mm., densely fimbriate; petals lanceolate-linear, 10–14 mm., 3(–5)-nerved, long woolly-fimbriate, inserted at the same level as or up to 5 mm. above the corona; filaments 3–5·5 mm., free or nearly so, inserted at the base of the hypanthium; anthers 8–12 mm., obtuse; septa 0–0·5 mm. high; corona consisting of hair-like threads 1·5–3 mm.; disk glands 1·5–3 mm. Female flowers broadly tubiform, including the 0·5–1·5 mm. long stipe 16–22(–25) mm.; hypanthium cup-shaped, 1·5–2 mm.; calyx-tube 7–10 by 8–11 mm., lobes oblong, obtusish, fimbriate, 5–8 mm.; petals linear, 8–12 mm., 1-nerved, ± fimbriate, inserted 1–2·5 mm. above the corona; staminodes 2·5–5 mm.; septa 0; corona of fine hairs ± 2 mm.; disk glands 1–2 mm.; gynophore 1–2·5 mm.; ovary ellipsoid, 4–5·5(–8) mm.; styles connate for 2–4 mm., style-arms ± 1 mm.; stigmas much branched, subreniform, ± woolly-papillate, each ± 4 mm. in diameter. Fruits 1 per inflorescence, subglobose to ellipsoid, excluding the 1–3 mm. long gynophore 3·5–5·5 by 3–4·5 cm.; pericarp red when ripe, coriaceous, inside ± spongy. Seeds 15–30 per capsule, ovate, 8–9 mm. Fig. 5/4, p. 21.

KENYA. Northern Frontier Province: Lerogi [Leroki] Plateau, 29 Sept. 1925, *Leakey* 32!; Nairobi District: Bahati, 17 Oct. 1932, *C.G. Rogers* 12!; Machakos District: Sultan Hamud, 20 Sept. 1953, *Drummond & Hemsley* 4426!
TANZANIA. Moshi District: Dutch Corner to Engare Nanyuki, 24 Oct. 1965, *Greenway & Vesey-FitzGerald* 12202! & Moshi, 25 Feb. 1953, *Drummond & Hemsley* 1323!; Mpwapwa District: Kongwa, Jan. 1967, *Wigg* in *E.A.H.* 13735!
DISTR. **K**1, 3–7; **T**1, 2, 4, 5; not known elsewhere
HAB. Grassland, deciduous and dry evergreen bushland; (0–)900–1750 m.

NOTE. *Verdcourt* 1603 bears male flowers as well as semi-mature fruits. The species is repeatedly reported as deadly poisonous, especially the roots. Closely related to 16, *A. ellenbeckii*; see note under that species.

19. **A. lanceolata** *Engl.* in E.J. 14: 378 (1891); Harms in E. & P. Pf. III. 6A: 84 (1893); Engl., V.E. 3(2): 603 (1921); F.P.S. 1: 163 (1950); de Wilde in Meded. Landbouwhogeschool Wageningen 71–18: 185 (1971) & in U.K.W.F.: 165 (1974). Types: Sudan, Bahr el Ghazal, Jur [Seriba Ghattas], *Schweinfurth* 1570, 1834 & 1837 (B, syn.!)

Suberect herb or climber to 5 m., glabrous, growing from a tuberous root-stock. Leaf-blades grey to glaucous, densely punctate or not beneath, entire, ovate or obovate to lanceolate, base attenuate to rounded, top acute-acuminate to broadly obtuse or subtruncate, sometimes with a subapical mucro, 1–15 by 0·5–5 cm., 3–5-nerved from near base to ± pinninerved; petiole 0·2–3 cm. Glands at blade base 1 or 2, in the latter case these contiguous or free, either situated on a wart-like or subspathulate median appendage, or on 2 ± connate or separate auricles; blade-glands 0, or 1–3 pairs ± approximate to the nerve-axils. Stipules narrowly triangular to linear, 1–1·5 mm. Inflorescences peduncled up to 2(–3) cm., 2–20(–50)-flowered in ♂, 1–3(–5)-flowered in ♀, tendril 0 or 1, 3–7(–10) cm.; sterile tendrils 2–10 cm. Male flowers tubular-infundibuliform, including the (2–)4–10 mm. long stipe (13–)15–27 mm.; hypanthium cup-shaped, 1·5–3·5 by 2–5 mm.; calyx-tube (4–)5–10 mm., lobes oblong to lanceolate, sub-obtuse, 5–8(–10) mm., serrulate-fimbriate; petals lanceolate, 5·5–9 mm., 3-nerved, serrulate, inserted at the same level as or up to 2 mm. above the corona; filaments 2–4·5 mm., connate for 0·5–2·5 mm., inserted at the base of the hypanthium or on a short androgynophore; anthers 3·5–5 mm., obtuse or subobtuse; septa 1–2·5 mm. high; corona hairs fine, ± 0·5–1 mm.; disk glands ± 1–1·5 mm. Female flowers tubular-campanulate, including the 1–7 mm. long stipe 10–22 mm.; hypanthium shortly cup-shaped, 1–3 by 3–5 mm.; calyx-tube 2·5–7·5 mm., up to 6 mm. wide; calyx-lobes oblong, obtuse to acute, 3–6(–8) mm., serrulate-laciniate; petals lanceolate or lanceolate-linear, 3–7·5 mm., 1–3-nerved, denticulate, inserted at the same level as the corona; staminodes 2·5–4 mm.; septa 0–1·5 mm. high; corona hairs 0·5–1 mm.; disk glands 0·5–1 mm.; gynophore 1·5–3 mm.; ovary ovoid-ellipsoid, 3·5–6 mm.; styles connate for 0·5–1·5 mm., style arms 0·5–1 mm.; stigmas subglobose, ± laciniate-papillate, each 1–2·5 mm. in diameter. Fruits 1–2 per inflorescence, ovoid-ellipsoid, excluding the 2–10 mm. long gynophore 2–4 by 1·5–2·5 cm.; pericarp coriaceous. Seeds 10–30 per capsule, ovoid to orbicular, 5–7 mm.

NOTE. This is a variable species, in habit as well as in size of flowers, fruits and seeds, and in the size and shape of the leaves. Two largely allopatric subspecies are recognized. In the abundant material a specimen with hermaphrodite flowers was found once; also incidentally flowers with a 6-merous perianth, or with 4 styles have been encountered.

The species has been recorded several times as poisonous. Various medicinal properties are ascribed to it.

subsp. **lanceolata**; de Wilde in Meded. Landbouwhogeschool Wageningen 71–18: 186 (1971)

Leaf-blades elliptic-oblong to lanceolate, 2–15 by 0·5–3·5 cm. Glands at blade-base 2, on 2 small separate auricles at the top of the petiole. Inflorescences up to 10(–15)-flowered in ♂, 1–3-flowered in ♀. Male flowers including the 4–10 mm. long stipe (13–)15–27 mm. Female flowers variable in size, including the 2–7 mm. long stipe 10–22 mm. Fruit excluding the 2–6 mm. long gynophore 2–4 cm.

UGANDA. Karamoja District: Kotido, July 1958, *J. Wilson* 480!; Teso District: Serere, Apr. 1932, *Chandler* 552! & Kyere, Feb. 1933, *Chandler* 1077!
TANZANIA. Mwanza District: Massanza I, 16 Mar. 1953, *Tanner* 1284!; Mpanda District: about 80 km. Mpanda–Ikola, 31 Oct. 1959, *Richards* 11677!; Lindi District: Kingupira, 28 Feb. 1971, *Ludanga* 1288!
DISTR. **U**1, 3; **T**1, 4, 8; S. Sudan
HAB. Deciduous woodland, bushland and thicket, wooded grassland, rocky hills; 900–1800 m.

SYN. *A. lanceolata* Engl. var. *grandifolia* Engl. in E.J. 14: 378 (1891). Type: Sudan, Bahr el Ghazal, Jur [Seriba Ghattas], *Schweinfurth* 109 (B, holo.†, K, iso.!)

NOTE. Very variable in habit, possibly partly due to burning. Also variable in the size of the flowers and fruits. Several specimens from the area east of Lake Tanganyika (**T**4) are very slender, and have remarkably small female flowers and fruits.

subsp. **scheffleri** (*Engl.*) *de Wilde* in Blumea 17: 180 (1969) & in Meded. Landbouw-hogeschool Wageningen 71–18: 187 (1971). Type: Kenya, Machakos District, Kibwezi, *Scheffler* 213 (B, holo.†, HBG, iso. !)

Leaf-blades ovate or obovate to lanceolate, 1–6(–8·5) by 0·5–2·5(–5) cm. Glands at blade-base 1 or 2, either single or contiguous on a median wart-like or subspathulate appendage, or on 2 ± connate auricles at the top of the petiole, making the blade slightly peltate. Inflorescences up to 50-flowered in ♂, 1–3(–5)-flowered in ♀. Male flowers including the 3–7·5 mm. long stipe 17–26 mm. Female flowers variable in size, including the 1–6 mm. long stipe 10–22 mm. Fruit excluding the 2–10 mm. long gynophore (2–)2·5–4 cm. Fig. 5/5, p. 21.

KENYA. Machakos District: 13 km. E. of Mtito Andei, 9 Nov. 1958, *Greenway* 9544 !; Kitui District: N. of Galana R., 8 Nov. 1965, *Hucks* 512 !; Teita District: Mbololo Plain, Oct. 1938, *Joana in Bally* 9083 !
TANZANIA. Mbulu District: Tarangire National Park, 23 Nov. 1968, *Richards* 23457 !; Dodoma District: Manyoni, 17 Dec. 1935, *B.D. Burtt* 5397 !; Morogoro District: 26 km. E. of Morogoro, 25 Oct. 1955, *Milne-Redhead & Taylor* 7378 !
DISTR. **K**4, 7; **T**1–3, 5, 6; Zambia, Malawi
HAB. Deciduous bushland; 450–1200 m.

SYN. *A. scheffleri* Engl., V.E. 3 (2): 603 (1921); Harms in N.B.G.B. 8: 294 (1923)

NOTE. Most specimens from Kenya and Tanzania have small female flowers, measuring 10–13(–15) mm.
    Specimens from Mpwapwa (**T**5), e.g. *Hornby* 551 and 640, have large leaves with subtruncate top, the female flowers 17–22 mm. long, and the fruits ± 4 cm. long, on 7–10 mm. long gynophores.

20. **A. digitata** (*Harv.*) *Engl.* in E.J. 14: 375 (1891) & V.E. 3(2): 605 (1921); Liebenberg in Bothalia 3: 527, 541, fig. 1–3, 14–17, t. 6–36 (1939); A. & R. Fernandes in Garcia de Orta 6: 259 (1958); de Wilde in Meded. Landbouwhogeschool Wageningen 71–18: 188, fig. 30, 31/a-h (1971). Type: South Africa, Natal, *Owen* (TCD, holo., K, photo. !)

Subherbaceous climber 0·2–3 m., growing from a variable tuber. Leaves beneath greyish to glaucous, densely punctate or not, deeply (3–)5-partite or pseudo-foliolate, in outline suborbicular with cordate base, 4–18 by 3–17 cm.; leaflets variable, entire to deeply (2–)3–5(–10)-lobed, ovate or obovate to linear, base long-attenuate or acute, top rounded to acute, 1·5–15 by 0·5–4(–7) cm., margin entire; petiolule up to 2 cm.; petiole 1–9 cm. Glands at blade-base 2, on 2 separate ± upward directed auricles 1–2 mm. in diameter at the transition of petiole to blade; blade glands 2–4, situated between the insertions of the leaflets, and each leaflet with 0–8 glands scattered or submarginal. Stipules narrowly triangular to lanceolate, acute, 1–3 mm., withering. Inflorescences peduncled for up to 7 cm., (1–)5–20(–60)-flowered in ♂, 1–10-flowered in ♀, tendril 1, 2–10 cm.; sterile tendrils up to 15 cm. Male flowers tubular-infundibuliform, including the 3–12(–15) mm. long stipe (14–)20–38 mm.; hypanthium cup-shaped, (1–)2–3·5 by 2–4(–5) mm.; calyx-tube (5–)8–12 mm.; calyx-lobes ovate or oblong to lanceolate, subobtuse, (4–)7–11 mm., dentate-fimbriate; petals lanceolate, acute, dentate or fimbriate, 6–12 mm., (1–)3-nerved, inserted at the same level as or up to 5 mm. above the corona; filaments 3·5–9(–12) mm., connate for at least half-way, inserted on an androgynophore 1–3·5 mm.; anthers 3–6 mm., ± inward curved and at the top clinging by the ± 0·2 mm. long papillate apicula; septa 0·5–3 mm. high, corona hairs ± 0·5–2 mm., rarely absent; disk glands 0·5–1·5 mm. Hermaphrodite flowers 20–25 mm. Female flowers tubular-infundibuliform, including the 2–7 mm. long stipe 15–26 mm.; hypanthium ± 2–4 by 2–4 mm.; calyx-tube 4–8 mm., lobes ovate to oblong, obtusish, 5–7 mm., entire; petals lanceolate to linear, acute, subentire, 2–7 mm., 1(–3)-nerved, inserted at the same level as or up to 4 mm. above the corona; staminodes 3–5 mm., up to 1 mm. connate; septa up to 0·5 mm. high; corona hairs 0·3–1 mm.; disk glands ± 1 mm.; gynophore 2–4 mm.; ovary ovoid to oblong, (4–)5–6 mm.; styles connate for 1–1·5 mm., style-arms 1–1·5 mm.; stigmas subreniform, woolly-papillate, each 2–3 mm. in diameter. Fruits 1–3 per inflorescence, ovoid to ellipsoid-oblong, excluding the (2–)5–12 mm. long gynophore (2·5–)3–5·5(–7·5) by (1·5–)2–3·5(–4) cm.; pericarp coriaceous, brilliant yellow to red when fresh, sometimes spongy inside, smooth. Seeds (10–)20–60 per capsule, obovoid to ellipsoid, 6–8 mm.

TANZANIA. Iringa District: Mufindi loop road, 17 Feb. 1961, *Richards* 15707!; Masasi District: W. of Bangala R., 16 Dec. 1955, *Milne-Redhead & Taylor* 7499!
DISTR. **T**7, 8; Angola (rare), Zambia (rare), Rhodesia, Malawi, Mozambique, Botswana, Bot. 1: 143 (1861) Swaziland, South Africa
HAB. Deciduous woodland and wooded grassland; 390–1800 m.

SYN. *Modecca digitata* Harv., Thes. Cap. 1: 8, t. 12 (1859) & in Fl. Cap 2: 500 (1862)
    Clemanthus senensis Klotzsch in Peters, Reise Mossamb., Bot. 1: 143 (1861). Type: Mozambique, Sena, *Peters* (B, holo.†, photograph in Liebenberg, loc. cit.)
    Modecca senensis (Klotzsch) Mast. in F.T.A. 2: 513 (1871): Hook. f. in Bot. Mag. 127, t. 7763 (1901)
    Adenia senensis (Klotzsch) Engl. in E.J. 14: 375 (1891); Harms in E. & P. Pf., ed. 2, 21: 491 (1925); A. & R. Fernandes in Garcia de Orta 6: 258 (1958)
    A. stenophylla Harms in E.J. 26: 238 (1899); Engl., V.E. 3 (2): 605 (1921). Type: South Africa, Transvaal, *Wilms* 941 (B, holo.†)
    A. multiflora Potts in Ann. Transv. Mus. 5: 235 (1917); Harms in N.B.G.B. 8: 298 (1923). Type: South Africa, *Potts* (*Fehrson*) in *Transv. Mus.* 17386 (PRE, holo.!)
    A. angustisecta Burtt Davy in K.B.: 280 (1921) & Man. Fl. Pl. & Ferns Transv.: 222 (1926). Type: South Africa, Transvaal, *Mundy* 4700 (BOL, holo.)
    A. buchananii Engl., V.E. 3 (2): 605 (1921). Type: Malawi, without precise locality, *Buchanan* 244 (K, lecto.!)

NOTE. The specimen *Milne-Redhead & Taylor* 7499, from Masasi District, **T**8, is deviating by its hermaphrodite flowers and may belong to the related species *A. kirkii* or may be a hybrid of *A. kirkii* and *A. digitata*. The locality of the specimen is in a transitional area between both species and the usually aberrant hermaphrodite flowers prevent a definite determination. More collecting in the area is needed.

21. **A. kirkii** (*Mast.*) *Engl.* in E.J. 14: 375 (1891); Harms in E. & P. Pf. III. 6A: 84 (1893); Engl., V.E. 3(2): 605 (1921); Watt & Breyer-Brandwijk, Medic. & Pois. Pl. S. & E. Afr., ed. 2: 828 (1962); de Wilde in Meded. Landbouwhogeschool Wageningen 71–18: 193, fig. 31/i, 32 (1971). Type: Tanzania, Dar es Salaam, *Kirk* (K, holo.!)

Herbaceous climber to 3 m. Leaves grey-glaucous beneath, punctate or not, 3–5-foliolate, suborbicular in outline, base ± cordate, 3–14 by 2·5–13 cm.; leaflets entire, or (especially the middle ones) pinnately up to 5-lobed, ovate to oblong, base acute, top obtuse to acute, 1–10 by 0·5–5(–7) cm., reticulation indistinct, margin entire; petiolule (0–)0·2–2·5 cm.; petiole 1–5 cm. Glands at blade-base 2, on 2 ± upward curved auricles 1–3 mm. in diameter at the transition to the petiole; blade-glands 2–4 between the insertions of the leaflets, and up to 10 small glands scattered or submarginal, rarely with 2 auricle-like glands at the top of the petiolule. Stipules linear, ± 1 mm., withering. Inflorescences pedunceled for 1–6 cm., up to 15-flowered in ♂, 1–4-flowered in ♀; tendril 2–6 cm.; sterile tendrils up to 12 cm. Male flowers tubular-infundibuliform, including the 15–25 mm. long stipe 25–40 mm.; hypanthium including calyx-tube (5–)7–9 mm., 2–4 mm. wide; calyx-lobes oblong-lanceolate, obtuse, 5–6·5 mm., subentire; petals lanceolate or oblanceolate, obtuse, (4–)5–6 mm., 3-nerved, minutely serrulate at the top, inserted (4–)5–6 mm. above the base of the hypanthium; filaments 4–5 mm., connate for ± 1·5 mm., inserted on an androgynophore ± 0·5 mm.; anthers 4·5–6 mm., obtuse, up to 0·1 mm.

apiculate; septa ± 1 mm. high; corona 0; disk glands 0·5–1 mm. Female flowers tubular-infundibuliform, including the 8–12(–15) mm. long stipe 16–24(–30) mm.; hypanthium including calyx-tube 2–6·5 mm., 2–4 mm. wide; calyx-lobes ovate to oblong, subobtuse, 4–5 mm., subentire; petals lanceolate to linear, subobtuse, 2·5–4·5 mm., (1–)3-nerved, subentire, inserted 1–3·5 mm. above the base of the hypanthium; staminodes 2–4 mm., connate for 0·5–1 mm., sometimes bearing abortive anthers, inserted at the base of the hypanthium; septa up to 1 mm. high; corona 0; disk glands ± 0·5 mm.; gynophore 2–3 mm.; ovary ovoid-ellipsoid, 3–3·5 mm.; styles connate for 0·5–1 mm., style-arms 0·5–1 mm.; stigmas subglobose, finely papillate-laciniate, 1·5–2 mm. in diameter. Fruits 1(–2) per inflorescence, ovoid to ellipsoid, excluding the ± 5 mm. long gynophore (2–)3–3·5 cm.; pericarp coriaceous. Seeds ± 40 per capsule, obovate to subrotund, ± 5–5·5 mm.

KENYA. Kilifi District: Lower Sabaki (Galana) R., May 1960, *Rawlins* 891! & Gida Forest, 8 Apr. 1947, *Jeffrey* K. 569!; Lamu District: Utwani, Dec. 1956, *Rawlins* 292!
TANZANIA. Mbulu District: Lake Manyara National Park, Jan. 1965, *Greenway & Kanuri* in *E.A.H.* 13096!; Morogoro plains, 1 Apr. 1932, *Wallace* 304!; Lindi District: 80 km. W. of Lindi, 23 Mar. 1935, *Schlieben* 6175!; Zanzibar I., 27 km. on Chwaka road, 2 Oct. 1961, *Faulkner* 2919!
DISTR. **K**7; **T**2, 3, 6, 8; **Z**; ? **P**; not known elsewhere
HAB. Shrub layer of lowland evergreen forest, grassland; 0–700 m.

SYN. *Modecca kirkii* Mast. in F.T.A. 2: 515 (1871)

NOTE. Characterized by the long, slender flower stipe. The species resembles 20, *A. digitata*, which has curved anthers clinging at their apices; in *A. kirkii* the anthers are straight and free at their tops.
    Leaves of juvenile specimens are sometimes variegated and have a distinct, up to 3 mm. deep peltate base.

22. A. trisecta (*Mast.*) *Engl.* in E.J. 14: 375 (1891); Hiern, Cat. Afr. Pl. Welw.: 384 (1898); Engl., V.E. 3(2): 605 (1921): A. & R. Fernandes in Garcia de Orta 6: 660 (1958); de Wilde in C.F.A. 4: 223 (1970) & in Meded. Landbouwhogeschool Wageningen 71–18: 195 (1971). Type: Angola, Cuanza Norte, Pungo Andongo, Zamba–Cazella, *Welwitsch* 863 (LISU, lecto.!, BM, iso.!)

Subherbaceous climber to ± 3 m., glabrous, growing from a tuberous rootstock. Leaves pale green, finely red-brown punctate beneath, 3(–5)-foliolate, ± suborbicular in outline, base subcordate, 4–8 by 5–10 cm., 3–5-nerved from base; leaflets entire, ovate to oblong, base attenuate, top acute to subobtuse, 2–6 cm.; margin entire; petiolule 0–1 cm.; petiole 1–4 cm. Glands at blade base 2, on 2 small auricles at the transition to the petiole; blade-glands 2(–4), situated in between the insertions of the leaflets. Stipules narrowly triangular, acute, 1–1·5 mm., withering. Inflorescences peduncled for 2–7·5 cm., up to 12-flowered in ♂, 1–3-flowered in ♀, tendril (0 or) 1, 3–6 cm.; sterile tendrils ± 8 cm. Male flowers ± tubiform-campanulate, including the 2–7 mm. long stipe 18–26 mm.; hypanthium cup-shaped, 2·5–4 mm.; calyx-tube 5–8 by 3–7 mm., lobes ovate-oblong, subacute, 5–9 mm., denticulate; petals lanceolate-linear, subacute, 5–6·5 mm., 3-nerved, denticulate-sinuate, inserted at the same level as or up to 2 mm. above the corona; filaments 2–2·5 mm., connate for ± 1 mm., inserted on a short androgyophore; anthers 5–5·5 mm., obtuse; septa 1·5–3 mm.; corona hairs sparse, 0·5–1 mm.; disk glands ± 1 mm. Female flowers campanulate, including the 1·5–2·5 mm. long stipe 10–15 mm.; hypanthium 1·5–2 mm.; calyx-tube 3·5–5·5 by 6–8 mm., lobes oblong, acute, 4–6 mm., entire; petals lanceolate-linear, subacute, 2–2·5 mm., 1-nerved, entire, inserted at the same level as or up to 0·5 mm. above the corona; staminodes 3 mm.; septa ± 1 mm. high; corona hairs sparse, ± 0·5 mm.; disk glands ± 0·5 mm.; gynophore 1·5–2 mm.; ovary ellipsoid, 4–5 mm.; styles connate for 0·5 mm., style-arms 2 mm.; stigmas woolly-papillate, each ± 2·5 mm. in diameter. Fruits 1–2 per inflorescence, sub-globose to ellipsoid, excluding the 2–3 mm. long gynophore ± 3 by 2.2 cm.; pericarp coriaceous. Seeds ± 30 per capsule, ovoid, 5–5·5 mm.

TANZANIA. Mpanda District: Mahali Mts., Lumbye R. mouth, 30 July 1958, *Newbould & Jefford* 1155!
DISTR. **T4**; Angola (Cuanza Norte)
HAB. Riverine forest, thicket, sandy soil; 800 m.

SYN. *Modecca trisecta* Mast. in F.T.A. 2: 514 (1871)

NOTE. Except the type collection from Angola, this species is only known from the second gathering from W. Tanzania. This specimen differs from the type by some-what broader male flowers, and by the small, not yet fully developed leaves subtending the inflorescences.
The species is related to *A. welwitschii* (Mast.) Engl. from Zaire and Angola.

23. A. goetzei *Harms* in E.J. 30: 360, t. 14 (1902); Engl., V.E. 3(2); 603 (1921); Harms in E. & P. Pf., ed. 2, 21: 491 (1925); de Wilde in Meded. Landbouwhogeschool Wageningen 71–18: 200 (1971). Type: Tanzania, Mbeya District, Toola, *Goetze* 1418 (B, holo.†)

Erect unbranched or sparingly branched herb up to 35 cm., without tendrils, glabrous, growing from a tuber. Leaf-blades somewhat glaucous, purple-brown spotted beneath, oblong-lanceolate to lanceolate-linear, base ± long-attenuate and inconspicuously auricled, top acute, 7–22 by (0·5–)1–6·5 cm., ± pinninerved; margin entire; petiole (0–)1–2(–4) mm. Glands at blade-base 2, partly situated on 2 ± inward curved auricles at the very base of the blade; blade-glands 0–20, scattered. Stipules narrowly triangular, entire or 2–5-dentate or -partite, 1–2·5 mm. Inflorescences sessile or up to 1 cm. peduncled, 1–5-flowered, the lower often axillary to 5–10 mm. long tricuspidate cataphylls; tendrils 0. Flowers polygamous. Male flowers resembling hermaphrodite flowers. Hermaphrodite flowers tubular-infundibuliform, including the 2·5–8 mm. long stipe 12–26 mm.; hypanthium (0·5–)1–3 mm.; calyx-tube 4–8(–10) by 3–8 mm., lobes oblong, obtuse 3–8(–9) mm., crispate-fimbriate; petals lanceolate-linear, acute, 5–12 mm., 1–3-nerved, fimbriate, inserted at the same level as or up to 3 mm. above the corona; filaments 3–7 mm., free or up to 2 mm. connate, inserted at the base of the hypanthium or on a short androgynophore; anthers 2·5–5 mm., sometimes abortive towards the top, provided with an up to 1 mm. long, mostly ± papillate apiculum; septa (0–)1–3 mm. high; corona hairs 1–2 mm.; disk glands ± 0·5 mm.; ovary in ♂ flowers vestigial; gynophore 2(–4) mm.; ovary ellipsoid-oblong, 3·5–5(–6) mm.; styles free or connate up to 1 mm., free style arms 1–1·5 mm.; stigmas subglobose, papillate, each 1–1·5 mm. in diameter. Female flowers resembling ♂ and hermaphrodite flowers, includ-ing the 2·5–5·5 mm. long stipe 16–22 mm.; petals 5–7 mm.; staminodes ± 2 mm.; septa 0–1 mm. high. Fruits 1 per inflorescence, pendent at maturity, ellipsoid to obovoid, excluding the curved 10–20 mm. long gyno-phore 3–4(–5) by 2–2·5 cm.; pericarp coriaceous. Seeds 30–40 per capsule, ovoid to subglobose, 4–4·5 mm.

TANZANIA. Kigoma District; 64 km. Ikola–Ulemba, 6 Nov. 1959, *Richards* 11723! & 11726!; Songea District: Songea Airfield, 1 Jan. 1956, *Milne-Redhead & Taylor* 7999! & Kwamponjore Valley, 7 Feb. 1956, *Milne-Redhead & Taylor* 7999B!
DISTR. **T4, 7, 8**; Zaire, Zambia, Rhodesia
HAB. *Brachystegia* woodland, bushland and wooded grassland; 900–1300 m.

Note. This species is usually polygamous. The flowers are variable in size and shape; relatively broad as well as narrow male flowers may be found on a single plant.

24. **A. bequaertii** *Robyns & Lawalrée* in B.J.B.B. 18: 284 (1947); de Wilde in Acta Bot. Neerl. 17: 131, fig. 1/b, 3 (1968) & in Meded. Landbouwhoge- school Wageningen 71–18; 243 (1971) & in U.K.W.F.: 164 (1974). Type: Zaire, Ruwenzori, *Bequaert* 3814 (BR, holo.!)

Somewhat woody climber to 20 m., up to 10 cm. thick at base; twigs grey-green or pruinose, not spotted. Leaf-blades ± grey-glaucous beneath, not punctate, entire or rarely ± 3-lobed in the upper half, ovate, base rounded to cordate, top acute, up to 1·5 cm. acuminate, 2·5–12 by 1·5–9 cm., 3(–5)-nerved from base with in addition 1(–2) pairs of nerves from the midrib not ending in the leaf-margin, reticulation distinct; margin entire or faintly toothed; petiole 2–10 cm. Gland at blade-base single on a ± convex subspathulate median appendage 1–3 mm.; blade-glands 2–5, distinct, 1(–2) in or close to the axils of the nerves; marginal glands minute, up to 12 on either side of the blade. Stipules broadly reniform, margin ± lacerate, 0·5–1(–1·5) mm. Inflorescences peduncled for (0·5–)2–16 cm., up to 40-flowered in ♂, 2–4-flowered in ♀, tendrils 1 or 3, 1–3 cm., or absent; sterile tendrils simple or 3-fid, 10–25 cm. Male flowers ± campanulate, including the 3–4 mm. long stipe 10–17 mm.; hypanthium shallowly cup-shaped, 1–1·5 by 2·5–4·5 mm., calyx-tube 0; sepals ± oblong-lanceolate, obtuse, subentire, (6–)8–11 by 2–3 mm., sparingly punctate or not; petals oblanceo- late, obtuse to acute, finely serrulate-laciniate, (6–)8–10(–11) mm., 3–5- nerved, ± punctate; filaments 2–3 mm., connate for 0·5–1·5 mm., inserted at the base of the hypanthium; anthers 4–5 mm., obtuse to subacute; septa ± 0; corona consisting of 5 cap-shaped parts alternating with the petals, 0·2–0·5 mm. high, sometimes superposed by a row of wart-like appendages or a knobbly rim; disk glands 0. Female flowers ± campanu- late, including the 0·5–4 mm. long stipe 7·5–12 mm.; hypanthium flattish, 0·5–0·8 by 2·5–3·5 mm.; calyx-tube 0; sepals oblong-lanceolate, obtuse to acute, serrulate towards apex, 5–7 mm., (1–)3-nerved; staminodes ± 1 mm.; septa 0; corona an inconspicuous rim ± interrupted by the petals; disk glands 0; gynophore ± 0·5 mm., ovary ovoid, 3–4 mm., finely punctate; style 1–1·5 mm.; stigmas subsessile, reniform to subglobose, papillate- laciniate, each ± 2 mm. in diameter. Fruits 1–3 per inflorescence, ovoid, top acutish, excluding the 1–2(–3) mm. long gynophore 3–5 by 1·7–2·5(–3) cm.; pericarp woody-coriaceous, ± 0·5 mm. thick, ± smooth to granulate. Seeds 50–60 per capsule, obliquely ovoid, 4–5 mm.

Uganda. Toro District: E. Ruwenzori, Bwamba, July 1940, *Eggeling* 3986!; Kigezi District; Kanungu, June 1939, *Purseglove* 814! & Aug. 1950, *Purseglove* 3481!
Kenya. Trans-Nzoia District: Kitale, Apr. 1969, *Tweedie* 3630!; Kericho District: Sambret–Timbilil, Sept. 1961, *Kerfoot* 2836! & S. of Kericho, 2 Dec. 1967, *Perdue & Kibuwa* 9227!
Distr. **U2**; **K3**, 5; E. Zaire, Rwanda, Burundi (see note)
Hab. Upland rain-forest or riverine forest and associated bushland; 1350–2300 m.

Note. The typical subspecies is described here. Subsp. *occidentalis* de Wilde occurs in lowland rain-forest in central and W. equatorial Africa; subsp. *macranthera* de Wilde is described from a single gathering from the Central African Republic.

25. **A. cissampeloides** *(Hook.) Harms* in E. & P. Pf. III. 6A, Nachtr. 1: 255 (1897) & ed. 2, 21: 490 (1925); Engl., V.E. 3(2): 602 (1921); F.W.T.A., ed. 2, 1: 202, fig. 80 (1954); de Wilde in Acta Bot. Neerl. 17: 132, fig. 2/i (1968) & in Meded. Landbouwhogeschool Wageningen 71–18: 246, fig. 39/e, h–i (1971) & in U.K.W.F.: 164 (1974). Type: Fernando Po, *Vogel* 162 (K, holo.!)

Somewhat woody climber up to 25 m. high, up to 10 cm. thick at base; twigs pale green to grey-green, often pruinose, spotted or not. Leaf-blades slightly to strongly punctate, often ± glaucous beneath, not lobed, suborbicular to bluntly 5-angular, base cordate to truncate, top subobtuse to rounded, sometimes retuse, 3–14 cm. across, 3(–5)-nerved from base, and with 1(–2) pairs of ± straight nerves from the midrib ending in marginal glands; reticulation only clearly visible in the coarser veins; margin entire; petiole 2–9 cm. Gland at blade-base single, on a median circular to spathulate appendage 1–3 mm.; blade-glands (0–)2–4, rather approximate to the axils of the upper main nerves; marginal glands 2–7 at either side of the blade. Stipules broadly rounded, sometimes lacerate, ± 0·5 mm. Inflorescences peduncled for 1·5–15(–20) cm., up to 20-flowered in ♂, 2–6-flowered in ♀, tendrils 0 or 1–3, 0·5–2(–4) cm.; sterile tendrils simple or 3-fid, 10–20 cm. Male flowers ± campanulate, including the 2–4·5 mm. long stipe (10–)12–15 mm.; hypanthium cup-shaped, ±1(–1·5) by 2–3 mm.; calyx-tube 0; sepals oblong-lanceolate, obtuse to subacute, (6–)7–9 by 2–3 mm., subentire, ± punctate; petals lanceolate or oblanceolate, obtuse to subacute, 7–10 mm., 3–5-nerved, fimbriate-laciniate in upper half, sparingly punctate or not; filaments 1·5–2·5 mm., free or up to half-way connate, inserted at the base of the hypanthium; anthers 3·5–5 mm., obtuse; septa 0–0·3 mm. high; corona 0 or consisting of 5 inconspicuous cap-shaped parts ± 0·2 mm. high; disk glands 0. Female flowers campanulate, including the 0·5–2 mm. long stipe 6–10 mm.; hypanthium flattish, 0·5–0·7 by 2–3 mm.; calyx-tube 0; sepals oblong-lanceolate, subacute, 5–8 mm., entire, punctate or not; petals lanceolate to linear, 3–6·5 mm., 1-nerved, subentire, punctate or not; staminodes 0·5–1 mm., ± connate at base; septa ± 0; corona an inconspicuous fleshy annulus; disk glands 0; gynophore 0·3–1 mm.; ovary broadly ovoid to ellipsoid, ± 3(–6)-ribbed or -angled, 3–6 mm., smooth or finely warty or punctate; style 0·5–1 mm., stigmas 3(–4), subsessile, subglobose or flattened, papillate, each 1–1·5 mm. in diameter. Fruits 1–3 per inflorescence, broadly ovoid to ellipsoid or ± fusiform, sometimes distinctly (3–)4–6(–8)-angular, excluding the 1–2 mm. long gynophore 1·5–3·5 by 1·2–2·5 cm.; pericarp coriaceous or woody, 0·5–1 mm. thick, smooth or inconspicuously warted or pitted, spotted or not. Seeds 30–60 per capsule, ovoid, 3–3·5 mm.

UGANDA. Bunyoro District: Kitoba, Apr. 1942, *Purseglove* 1215!; Mengo District: Kipayo, Apr. 1914, *Dummer* 746!
KENYA. N. Kavirondo District: N. of Kakamega, *Maas Geesteranus*!
TANZANIA. Kigoma District: Ititie, R. Lugufu, 24 Dec. 1963, *Azuma* (*Kyoto Univ. Exped.*) 1036! & 80 km. S. of Kigoma, 17 Feb. 1964, *Itani* 22!; Morogoro District: Turiani–Mhonda Mission, 18 June 1957, *Semsei* 2663! (identification somewhat doubtful)
DISTR. **U**2, 4; **K**5; **T**4, ?6; W. and equatorial Africa
HAB. Rain-forest, clearings and edges; 900–1200 m.

SYN. *Modecca cissampeloides* Hook., Niger Fl.: 365 (1849)
Passiflora marmorea Linden, Cat., No. 8 (1853); Gard. Chron.: 235, fig. 51 (1871). Type not indicated
Ophiocaulon cissampeloides (Hook.) Mast. in F.T.A. 2: 518 (1871)
O. rowlandii Bak. in K.B.: 16 (1895). Type: Nigeria, *Rowland* (K, holo.!)
Adenia rowlandii (Bak.) Harms in E. & P.Pf. III.6A, Nachtr. 1: 255 (1897)
A. triloba Engl., V.E. 3 (2): 602 (1921); Harms in N.B.G.B. 8: 293 (1923); F.W.T.A., ed. 2, 1: 202 (1954). Type: Togo, *Schröder* 168 (B, holo.†)

NOTE. In *A. cissampeloides* the leaves are never lobed, contrary to most of the related species. In East African specimens the blade-glands are sometimes absent.
    The specimen *Semsei* 2663 from **T**6 is rather doubtful, and may represent *A. gummifera*.

26. **A. gracilis** *Harms* in E. & P. Pf. III. 6A, Nachtr. 1 : 255 (1897); Engl., V.E. 3(2) : 602 (1921); F.W.T.A., ed. 2, 1 : 202 (1954); de Wilde in Acta Bot. Neerl. 17 : 129, 132, fig. 2/k (1968) & in Meded. Landbouwhogeschool Wageningen 71–18 : 255 (1971). Types : Cameroun, Yaoundé, *Zenker & Staudt* 383 (B, syn.†, BM, K, isosyn. !) & 457 (B, syn.†, BM, BR, K, S, isosyn. !)

Subherbaceous climber up to 10 m., twigs brownish green or grey-green, sometimes pruinose, faintly spotted or not. Leaf-blades pale green to grey-glaucous beneath, mostly warty-punctate, entire or up to half-way 3–5(–7)-lobed, orbicular to ovate or ± 3–5-angular in outline, base cordate to rounded, top obtuse to acute-acuminate, 1–6(–7·5) by 1–5·5(–8) cm., 3-nerved from base and in addition (1–)2 pairs of nerves from near the base of the midrib, ascending or ± straight with sharp angles from the midrib and ending in marginal glands; reticulation fine, usually distinct; margin entire or minutely toothed; lobes triangular to rounded, up to 2 cm.; petiole 1–7 cm. Gland at blade-base single, on a median subcircular to spathulate appendage 1–3 mm.; blade glands 0–6, submarginal or scattered; marginal glands minute, mostly tooth-like, blackish, 1–6 at either side of the blade. Stipules broadly rounded, finely lacerate, 0·5(–1) mm. Inflorescences peduncled for (0·2–)0·5–1·5(–3) cm., up to 10-flowered in ♂, 1–3-flowered in ♀, usually in the axils of smaller leaves of special inflorescence-bearing twigs, tendril 0 or 1, 0·5–1·5 cm.; sterile tendrils simple or 3-fid, 6–15 cm. Male flowers ± campanulate, including the 1·5–5 mm. long stipe 9–13(–15) mm.; hypanthium shallowly cup-shaped, 0·7–1·5 by 2–3 mm.; calyx-tube 0; sepals lanceolate, subobtuse to subacute, 5–8 by 1·5–2·5 mm., subentire, punctate or dotted; petals oblanceolate to linear, obtuse, 5–8·5 mm., 3(–5)-nerved, lacerate-fimbriate in upper half, sparingly punctate or not; filaments 2–2·5(–3) mm., connate for about 1–1·5 mm., inserted at the base of the hypanthium; anthers 3·5–5 mm., obtuse; septa up to 0·3 mm. high; corona conspicuous, either as a 5-partite fleshy annulus or as 5 cap-shaped parts, 0·2–0·5 mm. high; disk glands 0. Female flowers campanulate, including the 0·2(–0·5) mm. long stipe 3·5–4·5 mm.; hypanthium flattish, 0·3–0·5 by 1·5–2 mm.; calyx-tube 0; sepals oblong-lanceolate, subacute, 2·5–4 by ± 0·7 mm., entire, sparingly punctate; petals linear, subacute, ± 1 mm., 1-nerved, subentire, not punctate; staminodes ± 0·3 mm.; septa ± 0·2 mm. high; corona a fleshy 5-lobed annulus, ± 0·3 mm. high; disk glands 0; gynophore 0·3–0·5 mm.; ovary ovoid, 2–3 mm.; style 0·7–1 mm.; stigmas subsessile, subreniform, laciniate-papillate, each ± 0·7 mm. in diameter. Fruits 1(–2) per inflorescence, ovoid to ellipsoid-oblong, ± fusiform, faintly 3(–6)-angular, excluding the 0·5–1 mm. long gynophore 1–2 cm.; pericarp coriaceous, 0·2–0·5 mm. thick, smooth or finely pitted or warted. Seeds 25–40 per capsule, subovoid, 3–4 mm.

UGANDA. Mengo District: Kipayo, June 1914, *Dummer* 860 !
TANZANIA. Kigoma District: Gombe Stream National Park, 23 Feb. 1970, *Clutton-Brock* 430 !
DISTR. U4; T4; W. and central Africa, from Nigeria to Angola, east to Zaire (see note)
HAB. Forest edges; 1200 m.

SYN. *Ophiocaulon gracile* (Harms) Pellegr. in Mém. Soc. Linn. Norm. 26 (2) : 123 (1924)

NOTE. The typical subspecies occurs in East Africa; subsp. *pinnata* de Wilde occurs in W. Africa, from Senegal to Nigeria, and is distinguished by more pinnately nerved leaves and larger fruits.

27. **A. gummifera** (*Harv.*) *Harms* in E. & P. Pf. III. 6A, Nachtr. 1; 255 (1897) & ed. 2, 21 : 490 (1925); Engl., V.E. 3(2) : 602 (1921); Liebenberg in Bothalia 3 : 523, 532, 535, fig. 10, 11 (1939); T.T.C.L. : 447 (1949); A. & R. Fernandes in Garcia de Orta 6 : 256 (1958); F.F.N.R. : 267 (1962); de Wilde

in Acta Bot. Neerl. 17:131, fig. 2/h (1968) & in Meded. Landbouwhogeschool Wageningen 71–18: 261 (1971) & in U.K.W.F.: 164 (1974).   Type: South Africa, Natal, *Drège* 5211 (P, S, iso.!)

Somewhat woody climber up to 30 m., up to 10 cm. thick at base; twigs green or grey-green, often pruinose.   Leaf-blades rarely punctate, entire to deeply 3(–5)-lobed, orbicular to ovate or rhomboid, or ± 3(–5)-angular in outline, base cordate to truncate, top obtuse or retuse, rarely subacute, 2·5–11 by 2·5–11 cm., 3-nerved from base with 1 pair of straight nerves from the midrib ending in marginal glands; reticulation rather fine, distinct; margin entire; lobes obtuse, up to 4 cm.; petiole 1·5–11 cm.   Gland at blade-base single, on a median subcircular to spathulate appendage 1–3 mm.; blade-glands 0–4, often 2 glands rather approximate to the axils of or contiguous with the upper side nerves; marginal glands 3–7 on either side of the blade.   Stipules broadly rounded to triangular, finely lacerate, 0·5(–1) mm. Inflorescences peduncled for (0·5–)1–12(–16) cm., up to 35-flowered in ♂, 2–6-flowered in ♀, tendril 0 or 1, 1–4 cm.; sterile tendrils simple or 3-fid, 5–20 cm.   Male flowers ± campanulate, including the 2·6 mm. long stipe 11–17 mm.; hypanthium cup-shaped, 1–2(–2·5) by 2–4 mm.; calyx-tube 0; sepals lanceolate, subobtuse, (7–)8–10 by 2–3 mm., punctate, margin up to 0·2 mm. laciniate; petals lanceolate or oblanceolate, obtuse, (6–)8–11 mm., 3-nerved, finely laciniate-serrulate, remotely punctate; filaments (1–)2–3·5 mm., connate for (0–)0·5–1·5(–2) mm., inserted at the base of the hypanthium; anthers 3–6 mm., obtuse, up to 0·1 mm. apiculate; septa 0–0·3 mm. high; corona 0; disk glands 0.   Female flowers ± campanulate, including the ± 0·5 mm. long stipe 5·5–8 mm.; hypanthium flattish, ± 0·5 by 2–2·5(–3) mm.; calyx-tube 0; sepals oblong, ± 4–6·5 mm., entire, punctate; petals lanceolate-linear, 2–4·5 mm., 1–3-nerved, subentire, sparingly punctate or not; staminodes ± 0·5 mm.; septa 0; corona 0; disk glands 0; gynophore ± 0·5 mm., ovary ovoid, 3–4·5 mm., style 0–0·5 mm., stigmas subsessile, subreniform, laciniate-papillate, each ± 1–1·5 mm. in diameter. Fruits 1–4 per inflorescence, ovoid to ellipsoid, sometimes ± 3(–6)-angular, excluding the ± 1 mm. long gynophore 2·5–4(–4·5) by 1·7–3 cm.; pericarp woody-coriaceous, ± 0·3 mm. thick, smooth or finely pitted or granulate. Seeds 30–50 per capsule, subovoid, 3·5–5·5 mm.   Fig. 5/6, p. 21.

UGANDA.   Karamoja District: Moroto, 1959, *J. Wilson* 741!; Mbale District: W. Bugwe Forest Reserve, 28 Sept. 1953, *Drummond & Hemsley* 4494!; Mengo District: 6·5 km. [Kampala–]Entebbe, Jan. 1932, *Lab. Staff* in *Snowden* 2339!
KENYA.   Kiambu/Nairobi District: Karura Forest, Mar. 1934, *Napier* 3205 in *C.M.* 5976!; Central Kavirondo District: Port Victoria, 27 Oct. 1948, *Glasgow* 48/3!; Kilifi District: Mida, *R.M. Graham* A.494 in *F.D.* 1873!
TANZANIA.   Moshi District: Himo, Jan. 1894, *Volkens* 1728!; NW. Uluguru Mts., 5 Dec. 1932, *Schlieben* 3049!; Songea District: 12 km. W. of Songea, 7 Jan. 1956, *Milne-Redhead & Taylor* 8075!; Zanzibar I., Kidichi, 14 July 1961, *Faulkner* 2874!
DISTR.   U1, 3, 4; K4, 5, 7; T1–3, 5–8; Z; eastern Africa from S. Ethopia to South Africa, west to E. Zaire, also Seychelles
HAB.   Forest and bushland of various types; 0–1700 m.

SYN.   *Modecca gummifera* Harv. in Fl. Cap. 2: 500 (1862)
       Ophiocaulon gummifer Mast. in F.T.A. 2: 518 (1871)
       [*O. cissampeloides* sensu Bak., Fl. Maurit. & Seych.: 106 (1877) & in J.L.S. 40: 74 (1910), *non* (Hook.) Mast.]
       Adenia rhodesica Suesseng. in Trans. Rhodes. Sci. Assoc. 43: 13 (1951).   Type: Rhodesia Marandellas, *Dehn* 696/52 (M, holo.!, BR, SRGH, iso.!)
       A. sp. 1 sensu F. White, F.F.N.R.: 268 (1962)

NOTE.   The var. *cerifera* de Wilde is endemic in Zambia, and occurs rather close to the Tanzanian border.   This variety is distinguished by a number of small morphological characters, and its glaucous appearance.

28. **A. reticulata** (*De Wild. & Th. Dur.*) *Engl.*, V.E. 3(2): 602 (1921); de Wilde in Acta Bot. Neerl. 17: 129, 123, fig. 1/j (1968) & in C.F.A. 4: 224 (1970) & in Meded. Landbouwhogeschool Wageningen 71–18: 267 (1971). Type: Zaire, Mbandaka [Coquillatville], *Dewèvre* 691a (BR, holo.!)

Somewhat woody climber up to 15 m., up to 10 cm. thick at base, older stems ± triangular in cross-section; twigs grey-green, sometimes pruinose, spotted or not. Leaf-blades often variegated, very pale beneath, punctate or not, entire to 3–5-lobed or with sinuate margin, broadly ovate to ± triangular or rhombic, base cordate to rounded or truncate, top acute, up to 2 cm. acuminate, sometimes 1 mm. mucronate, 2–11 by 1·5–10 cm., 3–5-nerved from base and with (0–)1–2(–4) pairs of nerves from the midrib; nerves ascending or straight, ending in the leaf margin or not, reticulation distinct or not; margin entire; lobes subtriangular or rounded, up to 2 cm.; petiole 1–6·5 cm. Gland at blade-base single, on a median circular to spathulate appendage 1–2·5 mm.; blade-glands 0–30, small, scattered or submarginal, sometimes ± axillary to the nerves; marginal glands minute, blackish, 2–7 at either side of the blade. Stipules broad, rounded, finely lacerate, ± 0·5 mm. Inflorescences peduncled for 0·5–5 cm., up to 20-flowered in ♂, 2–3(–5)-flowered in ♀, tendril 0 or 1, 0·5–1·5 cm.; sterile tendrils simple or 3-fid, 10–20 cm. Male flowers ± campanulate, including the 1·5–4·5 mm. long stipe 7–14 mm.; hypanthium cup-shaped, 0·5–1(–2) by 3·5 mm.; calyx-tube 0(–1) mm.; sepals oblong-lanceolate, obtuse to subacute, subentire, 5–8 by 2–3 mm., punctate; petals oblanceolate, obtuse, with fimbriate margin, 5·5–8·5 mm., 3(–5)-nerved, sparingly punctate; filaments 1·5–2 mm., connate for 0·5–1 mm., inserted at the base of the hypanthium; anthers 3·5–4 mm., obtuse; septa (0·2–)0·5–1 mm. high; corona 0; disk glands 0. Female flowers ± campanulate, including the 0·2–1 mm. long stipe 6–7·5 mm.; hypanthium shallowly cup-shaped, 0·5–1 by 3–3·5 mm.; calyx-tube 0; sepals oblong to lanceolate, subacute, 4·5–5·5 mm., subentire, punctate; petals lanceolate-linear, subobtuse to acute, 2·5–3 mm., 1-nerved, subentire, not punctate; staminodes ± 0·5 mm., connate for 0–0·3 mm.; septa 0–0·3 mm.; corona a low 5-lobed ring, ± 0·3 mm. high; disk glands 0; gynophore ± 0·5 mm.; ovary ovoid, ± 3-ribbed, 3–4·5 mm., smooth or finely warty or granulate, style (0–)0·7 mm., stigmas sessile, reniform to globose, papillate, each 1–1·5 mm. in diameter. Fruits 1(–2) per inflorescence, subglobose, ovoid or ellipsoid, sometimes ± 3(–6)-angular, excluding the 0·5–1·5 mm. long gynophore 2–3(–3·5) by 1·5–2(–2·5) cm.; pericarp coriaceous or woody, (0·5–)1–2 mm. thick, strongly warted to nearly smooth, or pitted. Seeds 30–50 per capsule, ovoid, 4–4·5(–5) mm. Fig. 5/7, p. 21.

UGANDA. Masaka District: NW. side of Lake Nabugabo, 9 Oct. 1953, *Drummond & Hemsley* 4713!; Mengo District: Kipayo, May 1914, *Dummer* 794! & Kajansi Forest, Apr. 1935, *Chandler* 1248!
TANZANIA. Lushoto District: Monga, *Zimmermann* in Herb. Amani 6581!; Kigoma District: Kasakati, Aug. 1965, *Suzuki* 286! & 241!
DISTR. **U**4; **T**3, 4; W. and central Africa, from S. Nigeria to N. Angola
HAB. Rain-forest, semi-swamp and riverine forest; 900–1200 m.

SYN. *Ophiocaulon reticulatum* De Wild. & Th. Dur. in Compt. Rend. Soc. Bot. Belg. 38: 86 (1899)
     Adenia lobulata Engl., V.E. 3 (2): 602 (1921); Harms in N.B.G.B. 8: 293 (1923). Types: Zaire, Mukenge, *Pogge* 949 & Cameroun, Douala, *Hückstädt* 44 (both B, syn.†)

NOTE. The East African specimens belong to the typical variety, as described above; in W. equatorial Africa var. *cinerea* de Wilde occurs also and is distinguished by a number of small characters.

29. **A. stolzii** *Harms* in F.R. 11 : 35 (1913); de Wilde in Acta Bot. Neerl.
17: 131, fig. 2/g (1968) & in Meded. Landbouwhogeschool Wageningen
71–18: 271 (1971) & in U.K.W.F.: 164 (1974). Type: Tanzania, Rungwe
District, Bundali Mt., *Stolz* 147 (B, holo.†, BM, K, L, M, S, W, Z, iso. !)

Somewhat woody climber up to 20 m.; twigs grey-green. Leaf-blades not
or sparingly punctate, entire or subentire, orbicular to ovate, base cordate,
top obtuse to acute, 4–12 by 3·5–10 cm., 3(–5)-nerved from base and with
1(–2) pairs of nerves from the midrib ascending towards the top; reticulation
distinct; margin entire or rarely 0·5 cm. sinuate; petiole 1·5–11 cm. Gland
at blade-base single, on a median spathulate to wart-like appendage
1·5–2·5 mm.; blade-glands (0–)2–4, rather approximate to the axils of the
upper lateral nerves; marginal glands up to 6 at either side of the blade.
Stipules reniform, finely lacerate, 0·5–1 mm. Inflorescences axillary to
normal leaves or sometimes in the axils of reduced leaves on short-shoots,
peduncles 0·2–5 cm., up to 15-flowered in ♂, 1–3-flowered in ♀, tendril 0 or 1,
1–2 cm.; sterile tendrils simple, 10–25 cm. Male flowers campanulate,
including the 2–3 mm. long stipe 12–15 mm.; hypanthium shallowly cup-
shaped, 1–1·5 by 2–4(–5) mm.; calyx-tube 0; sepals lanceolate, obtuse,
8·5–13 by 2–3 mm., subentire, remotely punctate; petals lanceolate or
oblanceolate, obtuse, 8–13 mm., 3–5-nerved, finely serrulate in upper half,
remotely punctate; filaments 2–3 mm., connate for 1–2 mm., inserted at the
base of the hypanthium; anthers 4·5–5 mm., obtuse; septa 0; corona 0 or
consisting of a low ring, ± 0·1 mm.; disk glands 0. Female flowers campanulate, including the ± 0·5 mm. long stipe ± 10 mm.; hypanthium
flattish, ± 1 by 4 mm.; calyx-tube 0; sepals lanceolate, subobtuse, 8–9 mm.,
subentire; petals linear, subacute, ± 3·5 mm., 1-nerved, entire; staminodes
± 0·5 mm.; septa 0; corona 0; disk glands 0; gynophore ± 0·5 mm.; ovary
ovoid-ellipsoid, ± 5·5 mm.; style 0–0·5 mm.; stigmas sessile, reniform with
papillate-laciniate margins, each ± 1·5 mm. in diameter. Fruits 1(–2) per
inflorescence, ovoid-oblong, subfusiform, excluding the 1–1·5 mm. long
gynophore ± 4(–4·5) by 2–2·5 cm.; pericarp coriaceous, 0·2–0·5 mm. thick,
smooth. Seeds ± 30 per capsule, ± ellipsoid, 4–5 mm.

KENYA. Meru District: between Meru and Nithi [Niti], 25 Feb. 1922, *Fries* 1958 !
TANZANIA. Ufipa District: Nsanga Forest, 8 Aug. 1960, *Richards* 13005 !; Iringa
District: Mufindi, 4 Oct. 1968, *Paget-Wilkes* 188 !; Rungwe District: Kyimbila, 25
Nov. 1907, *Stolz* 147 ! & 16 Aug. 1913, *Stolz* 2131 !
DISTR. **K**4; **T**4, 7; N. Malawi
HAB. Upland rain-forest and associated bushland; 1300–2000 m.

NOTE. The only specimen cited from Kenya, from Aberdare Mts. (*Fries* 1958, in UPS),
is sterile, and its identification is somewhat doubtful; its leaves recall certain forms of
A. gummifera

30. **A. tricostata** *de Wilde* in Acta Bot. Neerl. 17: 133, 127, fig. 1/f, 3
(1968) & in Meded. Landbouwhogeschool Wageningen 71–18: 273, fig. 44
(1971). Type: Cameroun, Bertoua, *Breteler* 1465 (WAG, holo. !)

Slender climber to 8 m.; twigs ± pale green, not punctate. Leaf-blades
not punctate, entire, broadly ovate to oblong, base cordate to subacute, top
acute, up to 2·5 cm. acuminate, (2·5–)4–14 by (1·5–)2–7·5 cm., 3-nerved
from base, nerves arching towards the top, with distinctly parallel venation
between; margin with minute gland-teeth; petiole 1–7 cm. Gland at blade-
base single, situated at the base of a median spathulate appendage 1–3·5
mm.; blade-glands absent; marginal glands minute, 0–10 at either side of
the blade. Stipules reniform with crenulate-lacerate edge, 0·5–1·5 mm.
Inflorescences peduncled for 3–15 cm., lax, up to 30-flowered in ♂, 1–4-
flowered in ♀, tendrils (0–)1 or 3, 0·5–2 cm.; sterile tendrils simple or 3-fid,

8–16 cm. Male flowers tubular-campanulate, including the 2·5–4 mm. long stipe (12–)15–21 mm.; hypanthium cup-shaped, 0·8–2·5 by 2–3·5(–4) mm.; calyx-tube 4–8 mm.; calyx-lobes lanceolate, subacute, subentire, 6–8 mm., not punctate; petals inserted in or up to 2·5 mm. below the throat of the calyx-tube, lanceolate-linear, acute, 6–10 mm., 1-nerved, minutely serrulate near the top; filaments 2–3 mm., connate for 1·5–2·2 mm., inserted at the base of the hypanthium; anthers 7·5–11 mm., obtuse; septa 0; corona consisting of a fleshy rim 0·5(–1) mm., superposed by an up to 0·5 mm. high 5-lobed or -hooded undulate rim; disk glands 0. Female flowers ± tubular-campanulate, including the 1–1·5 mm. long stipe 8–15 mm.; hypanthium cup-shaped, ± 1·5 by 2·5–3 mm.; calyx-tube 0–2 mm.; calyx-lobes (or sepals) lanceolate, obtuse to acute, entire, 5–11 mm., not punctate; petals inserted in or up to 1·5 mm. below the throat of the calyx-tube, linear, 3–5 mm., 1-nerved, entire; staminodes ± 1 mm., ± connate at base; septa 0–0·3 mm.; corona consisting of a low rim, superposed by a fleshy 5-lobed or -hooded ± undulated upper rim, or as a single fleshy rim up to 1 mm. high; disk glands 0; gynophore 0·5–2·5 mm.; ovary ovoid-oblong, 3·5–4·5 mm., smooth; style single 2–2·5 mm., stigmas ± reniform, laciniate-papillate, each ± 1·5 mm. in diameter. Fruits 1(–2) per inflorescence, not known at maturity, ovoid-ellipsoid, with acutish top, excluding the ± 6 mm. long stipe at least 2·5 cm. Seeds not known.

UGANDA. Mengo District: Nabugulo Forest, Bajo, Nov. 1916, *Dummer* 2996! & Kajansi Forest, May 1935, *Chandler* 1245! & May 1937, *Chandler* 1641!
DISTR. **U4**; Cameroun, Central African Republic, Congo (Brazzaville), Zaire
HAB. Rain-forest, often at edges, and riverine forest; 1050–1200 m.

6. **BASANANTHE**

Peyr. in Bot. Zeit. 17: 101 (1859) & in Wawra & Peyr., Sertum Benguelense, in Sitz. Ber. Acad. Wien 38: 569 (1860); Hook f. in G.P. 1: 812 (1867); Welw. in Trans. Linn. Soc. 27: 27 (1871); de Wilde in Blumea 21: 327 (1973)

Tryphostemma Harv., Thes. Cap. 1: 32, t. 51 (1859) & in Fl. Cap. 2: 499 (1862); Hook. f. in G.P. 1: 811 (1867); Engl. in E.J. 14: 387 (1891) & V.E. 3(2): 598 (1921); Harms in E. & P. Pf. III. 6A: 80 (1895) & ed. 2, 21: 487 (1925); Hutch. & Pearce in K.B.: 257 (1921); G.F.P. 2: 371 (1967)

Carania Chiov., Fl. Somala 1: 175 (1929); G.F.P. 2: 372 (1967)

Annual or perennial herbs or small climbers, rarely shrubs, glabrous or hairy, with or without tendrils. Leaves lobed or not, sessile or petiolate; margin entire or usually dentate with small glandular teeth. Stipules small, linear. False stipules present in some species, developed from the supra-axillary bud. Tendrils axillary, replacing central flower of cyme, or absent. Inflorescences axillary, cymose, 1–3-flowered, sessile or peduncled; bracts and bracteoles small, linear, often forming an involucre. Flowers bisexual (sometimes functionally unisexual), campanulate, greenish. Stipe indistinctly articulate at base to the short pedicel. Hypanthium rather narrow, flattish, rarely shallowly cup-shaped. Sepals 5(–6), oblong to lanceolate, free. Petals absent, 1–2 or (4–)5(–6), oblong to lanceolate, obtuse or sub-obtuse, free, usually smaller than the sepals. Outer corona consisting of a ± barrel-shaped tube bearing a ring of filiform processes (threads), bluish and mostly with a ring of small inward curved teeth. Disk low, annular, rarely absent. Inner corona membranous, cup-shaped, margin entire or lobulate (in *B. berberoides* from Somalia forming 5 small cups around the bases of the filaments). Stamens 5(–6–9); filaments inserted in the upper half inside the inner corona, free; anthers basifixed, ellipsoid to lanceolate, subsagittate, 2-thecous. Ovary ellipsoid, superior, usually sessile, 1-locular, with 3(–4) placentas; styles 3(–4), free or partially united; stigmas globose, small. Fruit a sessile or shortly stiped 3(–4)-valved capsule, ellipsoid; valves coriaceous. Seeds 1-few, arillate, ellipsoid to reniform, ± compressed; testa coriaceous, mostly rugose, blackish.

A genus of 25 species in central, eastern and southern Africa.

As explained in my revision (loc. cit.) I have refrained from any subdivision of the genus.

False stipules are stipule-like appendages, much larger than the minute true stipules, developed from the supra-axillary serial bud or shoot.

Petals 0(–2):
  False stipules large, foliaceous . . . . 1. *B. apetala*
  False stipules absent or the first leaves (cataphylls) of supra-axillary shoot not distinctly appearing as false stipules:
    Leaves sessile or subsessile, unlobed; tendrils very rarely present . . . . 2. *B. sandersonii*
    Leaves petioled for (0·5–)1 cm. or more, lobed or not; tendrils present:
      Sepals ± 1·5 mm. long . . . . 3. *B. phaulantha*
      Sepals 5 mm. or more long . . . . 4. *B. zanzibarica*
Petals (4–)5(–6):
  Plant unarmed:
    Leaves sessile or petiolate, lobed or not, glabrous or hairy, rarely scabrous but if so then lobed and long-petiolate:
      Tendrils present; annual or perennial of various habit, (10–)20–360 cm. tall; leaves unlobed or deeply lobed:
        Leaves 3–5-lobed, rarely unlobed: petiole distinct, unwinged, without glands or with only a few filiform glands at top . 5. *B. hanningtoniana*
        Leaves unlobed or 3-lobed, sessile or petiolate; petiole winged, with gland-teeth . 6. *B. lanceolata*
      Tendrils absent; perennial with woody rootstock, shoots up to 40 cm.; leaves not or shallowly lobed:
        Plant entirely glabrous; leaves oblong to lanceolate . . . . . 6. *B. lanceolata*
        Plant hairy at least in part; leaves pubescent on one or both surfaces:
          Shoots erect; leaves elliptic to elliptic-oblong . . . . . 7. *B. pubiflora*
          Shoots prostrate; leaves suborbicular, shallowly lobed in the upper half . 8. *B. hederae*
    Leaves scabrous, subsessile, unlobed . . 9. *B. scabrifolia*
  Plant thorny . . . . 10. *B. spinosa*

1. **B. apetala** (*Bak. f.*) *de Wilde* in Blumea 21 : 338, fig. 1/e (1973). Types: Malawi, Zomba, *Whyte* (K, syn. !, BM, isosyn. !)

Erect perennial herb up to 70 cm., growing from a woody rootstock, glabrous. Tendrils up to 6 cm., sometimes absent. Leaf-blades elliptic to lanceolate, 1·5–8 by 0·5–2(–4·5) cm., base acute, apex obtuse to acute, mar-

gin serrate, the teeth to 1 mm., sharply mucronate; petiole 0·1–0·4(–0·6) cm.
Stipules 2–3 mm. False stipules asymmetrical, (0·5–)1–4 cm., acute at
both ends, margin serrate. Inflorescences 1–2-flowered; peduncle 1–5 cm.;
bracts 1·5–3 mm. Flowers glabrous; stipe 2–5 mm. Hypanthium ± 2·5 mm.
wide. Sepals 5, the inner ± petaloid, 4–5·5 mm., obtuse. Petals 0. Outer
corona-tube ± 2 mm.; threads 1–1·8 mm. Disk 0·2–0·5 mm. Inner corona
cup-shaped, 0·7–1 mm. Stamens 5; filaments 1·5–2·5 mm.; anthers 0·7–
1 mm. Ovary 0·7–1·3 mm.; styles 3(–4), 0·7–1·3 mm., free. Fruit (excluding
the 3–6 mm. long gynophore) 1·5–2 cm., containing 1–3 seeds. Seeds
± 7 mm.

Tanzania. Songea District: 54·5 km. W. of Songea, 28 Feb. 1956, *Milne-Redhead &
Taylor* 8759!
Distr. **T8**; Malawi, Rhodesia, Mozambique
Hab. *Brachystegia* woodland; 1050 m.

Syn. *Tryphostemma apetalum* Bak. f. in Trans. Linn. Soc., ser. 2, 4: 14, t. 3/7–11 (1894);
Hutch. & Pearce in K.B.: 263 (1921); A. & R. Fernandes in Garcia de Orta 6:
251 (1958)
T. apetalum Bak. f. var. *serratum* Bak. f. in J. B. 37: 437 (1899); Engl., V.E.
3 (2): 599, fig. 265 (1921); Harms in E. & P.Pf., ed. 2, 21: 487, fig. 222 (1925);
A. & R. Fernandes in Garcia de Orta 6: 251 (1958). Type: Rhodesia, Salisbury,
Marshall (BM, holo.!)

Note. The only specimen from the Flora region, *Milne-Redhead & Taylor* 8759, is a
deviating specimen with decumbent habit, rather broad leaves, and small but distinct
false stipules; it shows some resemblance in habit with the form of *B. sandersonii*
found in the same area.

2. **B. sandersonii** (*Harv.*) *de Wilde* in Blumea 21: 339, fig. 2/a–c (1973).
Types: South Africa, Durban [Port Natal], *Sanderson* 59 & 440 (TCD, syn.,
K, isosyn.!)

Perennial herb 2–60 cm., with 1–several shoots erect or ± prostrate at
base from a rootstock, glabrous. Tendrils sometimes present, 0·5–2(–5) cm.
Leaf-blades unlobed, suborbicular, broadly ovate (or obovate) or elliptic to
lanceolate, 2–16 by 0·6–4(–5·5) cm., base subcordate or rounded to acute,
sometimes attenuate, top broadly obtuse (to retuse) to mostly acute, margin
remotely serrate-dentate with teeth 0·5–1 mm., especially towards base;
petiole 0–0·3(–0·5) cm. Stipules 1·5–5 mm. False stipules absent. Inflorescences 1–3-flowered; peduncle up to 4·5 cm.; bracts 1–4 mm. Flowers
glabrous; stipe 3–17 mm. Hypanthium 2–4 mm. wide. Sepals 5–7 (the
inner ones petaloid), obtuse, 4·5–8·5 mm. Petals 0(–2). Outer corona-tube
1–2(–3) mm., threads 0·6–1·5(–2) mm., sometimes ± branched. Disk
0·1–0·5 mm. Inner corona cup-shaped, 1–1·5(–2·5) mm. Stamens 5;
filaments 1·5–3·5 mm.; anthers 0·6–1·3 mm. Ovary up to ± 1 mm. stiped,
1–1·5 mm.; styles 3–4, free, 3–4 mm. Fruit (excluding the 1–3 mm. long
gynophore) 1·2–2 cm., containing 1–4 seeds. Seeds 6–8 mm.

Tanzania. Songea District: Mbarangandu R., Dec. 1900, *Busse* 673! & Songea, 15
Jan. 1956, *Milne-Redhead & Taylor* 8249! & 13 Feb. 1956, *Milne-Redhead & Taylor*
8249A! & 18 Feb. 1956, *Milne-Redhead & Taylor* 8265!
Distr. **T8**; Mozambique, Rhodesia, Swaziland, South Africa
Hab. *Brachystegia* woodland; 900–1100 m.

Syn. *Tryphostemma sandersonii* Harv., Thes. Cap. 1: 33, t. 51 (1859) & in Fl. Cap. 2:
499 (1862); Hutch & Pearce in K.B.: 265 (1921)
T. natalense Mast. in Trans. Linn. Soc. 27: 639 (1871), *nom. illegit.* (a new name
for *T. sandersonii*)
T. longifolium Harms in E.J. 33: 149 (1902); Hutch. & Pearce in K.B.: 264
(1921); A. & R. Fernandes in Garcia de Orta 6: 252 (1958). Type: Tanzania,
Mbarangandu R., *Busse* 673 (B, holo.†, EA, iso.!)

T. viride Hutch. & Pearce in K.B.: 265 (1921). Types: South Africa, Transvaal, *Bolus* 7602, *Galpin* 931 & *Thorncroft* in *Wood* 4366 (K, syn!)
T. friesii Norlindh in Bot. Notis.: 107, fig. 9, 10 (1934); A. & R. Fernandes in Garcia de Orta 6: 251, 252 (1958). Type: Rhodesia, Inyanga, *Fries, Norlindh & Weimarck* 3112 (S, holo.!, BM, BR, PRE, iso.!)

NOTE. A very variable species in habit and leaf-shape, as well as in flower size.
   T. longifolium, from S. Tanzania, links up with narrow-leaved forms described as *T. viride* from South Africa. In the flowers of *T. longifolium* the outer corona-tube and the cup-shaped inner corona are relatively long, ± 3 and 1·5–2·5 mm. respectively. In these rather tall, narrow-leaved forms tendrils are sometimes present.

3. **B. phaulantha** (*Dandy*) *de Wilde* in Blumea 21: 341, fig. 2/e (1973).
Type: Tanzania, Mwanza, *Davis* 208 (K, holo.!)

Erect annual herb up to 60 cm., sometimes shortly branched at base, glabrous. Tendrils 1–3 cm. Leaf-blades deeply 3-lobed, up to 6 by 5 cm.; lobes elliptic to oblong, 1–5 by 0·7–2 cm., top obtuse to acute, margin 1–2 mm. deep serrate-mucronate, more densely and deeper towards the base; base cordate or subcordate; petiole (0·5–)1–4·5 cm. Stipules ± 2 mm. False stipules absent. Inflorescences (1–)2-flowered; peduncle up to 0·8 cm., bracts ± 1 mm. Flowers glabrous; stipe 1–2·5 mm. Hypanthium ± 0·7 mm. wide. Sepals 1·5–2 mm., subacute to obtuse. Petals 0. Outer corona-tube ± 1 mm.; threads ± 0·7 mm. Disk ± 0·1 mm. Inner corona cup-shaped, 0·2–0·4 mm. Stamens 5; filaments 0·5–0·7 mm.; anthers ± 0·5 mm. Ovary ± 0·7 mm.; styles ± 0·3 mm., free. Fruit (excluding the ± 2 mm. long gynophore) 6–7 mm., containing 1 seed. Seed 4–5 mm.

TANZANIA. Mwanza, 16 Apr. 1926, *Davis* 208!
DISTR. T1; Zambia, Malawi
HAB. Open places on clayish soils; 1150 m.

SYN. *Tryphostemma phaulanthum* Dandy in K.B.: 251 (1927)

4. **B. zanzibarica** (*Mast.*) *de Wilde* in Blumea 21: 342, fig. 2/f, g (1973).
Type: Tanzania, Dar es Salaam, *Kirk* (K, holo.!)

Climbing annual or perennial to ± 2 m., glabrous or pubescent (on lower leaf-surface and sepals). Tendrils 5–12 cm. Leaf-blades unlobed, broadly ovate to oblong, up to 10 by 6(–7) cm., or leaves deeply (2–)3-lobed, base rounded to cordate; lobes ovate to oblong, 1·5–7·5 by 0·7–3 cm.; top sub-acute; margin towards base with remote slender teeth 0·5–1 mm.; petiole 0·5–4 cm. Stipules 1·5–4 mm. False stipules absent. Inflorescences (1–)2-flowered; peduncle 1–6 cm. (rarely flowers sessile on short shoots); bracts 1–3 mm. Flowers glabrous or pubescent; stipe 3–7 mm. Hypanthium 1·5–3·5 mm. wide. Sepals (4·5–)6–7 mm., subacute. Petals 0. Outer corona-tube barrel-shaped, 2–2·5(–3) mm., threads ± 2 mm. Disk 0·2–0·5 mm. Inner corona cup-shaped, 1–1·5 mm. Stamens 5; filaments 3–3·5 mm.; anthers 1·5(–2) mm. Ovary (0·5–)1–2 mm.; styles 3(4), free or nearly so, (0·5–)2–3·5 mm. Fruit (excluding the 2–3 mm. long gynophore) ± 1·5 cm., containing 1–5 seeds. Seeds ± 5 mm.

KENYA. Kwale District: Buda Mafisini Forest, 22 Aug. 1953, *Drummond & Hemsley* 3956! & Buda Forest, 3 Nov. 1959, *Napper* 1373! & Shimba Hills, 14 Jan. 1964, *Verdcourt* 3924!
TANZANIA. Uzaramo District: Pugu Hills, 8 Jan. 1939, *Vaughan* 2724! & Vikindu Forest Reserve, Aug. 1953, *Semsei* 1327! & 17 km. S. of Dar es Salaam, 20 June 1971, *Wingfield* 1607!; Zanzibar I., Kidichi, 18 Dec. 1959, *Faulkner* 2436!
DISTR. K7, T6; Z; not known elsewhere
HAB. Lowland dry evergreen forest, coastal bushland; 0–450 m.

SYN. *Tryphostemma zanzibaricum* Mast. in F.T.A. 2: 508 (1871) & in Trans. Linn. Soc.
    27: 639 (1871); Engl. in E.J. 14: 388 (1891); Hutch. & Pearce in K.B.: 264
    (1921)
   T. pilosum Harms in P.O.A. C: 280 (1895); Hutch. & Pearce in K.B.: 264 (1921).
    Type: Tanzania, Uzaramo District, Vikindu, *Stuhlmann* 6127 (B, holo.†, BM,
    drawing of leaf)
   T. stuhlmannii Harms in P.O.A. C: 280 (1895). Type: Tanzania, Uzaramo
    District, Kisarawe, *Stuhlmann* 6243 (B, holo.†, BM, drawing of leaf)

NOTE. Resembles in habit certain forms of the *B. hanningtoniana* complex, with which
it possibly hybridizes.

5. **B. hanningtoniana** (*Mast.*) *de Wilde* in Blumea 21: 342, fig. 3/a–c
(1973). Types: Kenya, Teita District, Maungu, *Johnston* & Tanzania,
Morogoro District, Kwa Chiropa, *Hannington* (both K, syn. !)

Erect or climbing 0·1–3 m., branched, annual or perennial, glabrous,
pubescent, scabrous or hispid. Tendrils 1–15 cm. Leaf-blades 3–5-lobed,
rarely unlobed, ovate to suborbicular in outline, 1–13 by 1–13 cm., glabrous
or hairy (often only beneath), base truncate to cordate; lobes ovate or
elliptic to lanceolate, or obovate, up to 8·5 cm., top acute to broadly obtuse;
margin dentate or with remote slender gland-teeth to ± 3 mm., often more
dense towards the base; petiole 0·2–8 cm., sometimes with a few gland-
teeth at the top. Stipules 3–15 mm. False stipules absent. Inflorescences
(1–)2-flowered; peduncle 1–10 cm.; bracts 0·5–4 mm., sometimes caducous
or absent. Flower-stipe 2–10 mm. Hypanthium 1·5–4 mm. wide. Sepals
3–8 mm., obtuse to acute, glabrous or hairy. Petals 2–6 mm. Outer
corona-tube 1·5–2·3 mm., threads 1·5–2·3 mm. Disk 0·2–0·5 mm. Inner
corona cup-shaped, 0·7–1·5 mm. Stamens 5(–9); filaments 2–3 mm.,
anthers 1–1·5(–2·5) mm. Ovary 0·6–2 mm.; styles 3(–4), free, 1–3 mm.
Fruit (excluding the 2–6 mm. long gynophore) 1–1·8 cm., containing 1–5
seeds. Seeds 6–7 mm. Fig. 8.

UGANDA. Mbale District: Budadiri, Jan. 1932, *Chandler* 417!; Masaka District:
  Bugabo, Sept. 1963, *Tallantire* 615!; Mengo District: Kiwala, Jan. 1917, *Dummer*
  3047!
KENYA. Northern Frontier Province: Laisamis, Dec. 1956, *J. Adamson* 608!; Machakos
  District: near Kiboko, 9 Jan. 1964, *Verdcourt* 3861!; Lamu District: Witu, Feb. 1957,
  Rawlins 340!
TANZANIA. Lushoto District: World's View, 4 June 1953, *Drummond & Hemsley*
  2842!; Dodoma District: Imagi Hill, 28 Jan. 1962, *Polhill & Paulo* 1281!; Ulanga
  District: Mahenge, 28 Dec. 1931, *Schlieben* 1577!
DISTR. U3, 4; K1, 4, 7; T1–7; Sudan, Ethiopia, ? Somali Republic, Zaire, Zambia,
  Rhodesia, Malawi, Mozambique
HAB. Wide variety of open habitats, from forest edges to deciduous bushland, the
  delicate annual forms usually on poor sandy soils; (0–)200–2150 m.

SYN. *Tryphostemma hanningtonianum* Mast. in Hook., Ic. Pl., t. 1484 (1885); Engl. in
    E.J. 14: 390 (1891); Hutch. & Pearce in K.B.: 261 (1921); U.K.W.F.: 166
    (1974)
   T. niloticum Engl. in E.J. 14: 389 (1891) & 15: 577 (1893); Hutch. & Pearce in
    K.B.: 261 (1921). Type: Sudan, *Baker* 214 in *Herb. Schweinfurth* (B, holo.†,
    BM, drawing of leaf)
   T. volkensii Harms in E.J. 19, Beibl. 47: 40 (1894); Hutch. & Pearce in K.B.:
    261 (1921). Type: Tanzania, Moshi District, Marangu, *Volkens* 1485 (B,
    holo.†, BM, drawing of leaf)
   T. hanningtonianum Mast. var. *latilobum* Harms in P.O.A. C: 280 (1895), as
    " *latiloba* ". Type: Tanzania, Lushoto District, Derema [Nderema], *Volkens*
    117 (B, holo.†, BM, iso. !)
   T. latilobum (Harms) Engl., V.E. 3 (2): 599 (1921)
   T. snowdenii Hutch. & Pearce in K.B.: 261, fig. 1 (1921). Type: Kenya, Kiambu
    District, Limuru, *Snowden* 598 (K, holo. !, BM, iso. !)
   T. stolzii Engl., V.E. 3 (2): 599 (1921); Harms in N.B.G.B. 8: 291 (1923). Types:
    Tanzania, Rungwe District, Kyimbila, *Stolz* 603 (B, syn.†, BM, K, L, isosyn. !)
    & Njombe District, Bulongwa, *Stolz* 2186 (B, syn.†)

FIG. 8. *BASANANTHE HANNINGTONIANA*—**1,** habit, × ⅔; **2,** lower surface of apex of middle lobe of leaf, × 4; **3,** flower, × 4; **4,** same, with sepals and petals removed, × 6; **5,** longitudinal section of flower, × 10; **6,** fruit, × 2; **7,** seed, × 4. 1–5, from *Drummond & Hemsley* 2842; 6, 7, from *Verdcourt* 3861. Drawn by Mrs. M. E. Church.

T. foetida Lebrun & Taton in B.J.B.B. **18**: 283 (1947). Type: Zaire, Kivu Province, Kasindi–Lubango, *Lebrun* 4715 (BR, holo. !)

T. sp. A sensu Agnew, U.K.W.F.: 166 (1974)

NOTE. This is a complex species which includes a number of more or less recognizable forms. Variability abounds in the growth form and habit. There are delicate erect annual forms, with weak roots, in which the cotyledons are persistent. These annuals have often relatively small flowers and fruits. Later on they may develop a climbing habit, becoming biennial or perennial. Coarse perennial forms are found mostly in mountainous areas (Usambara Mts.), and have a thick perennial rootstock. Often the flowers are variable in size in a single specimen, apparently partly in relation to their age. The indumentum varies as much. Glabrous forms are probably most common, but glabrescent or hairy forms are frequent as well; hairy specimens occur apparently only in the perennial forms. A few specimens (*Greenway & Kanuri* 11271, 11883, *Makin* 14036 and *Gillett* 19182) are remarkably scabrous. Leaf shape and texture is very variable. Unlobed leaves are rarely found.

A few specimens, viz. *Haerdi* 385/0, *Schlieben* 1577 and 2078, all from Ulanga District in Tanzania (**T**6), have some characters in common with *B. hanningtoniana* and 6, *B. lanceolata*, as well as with 4, *B. zanzibarica*. All have unlobed leaves, with distinct unwinged petioles with a few slender gland-teeth near the top. A toothed petiole is characteristic of certain forms of *B. lanceolata*. In habit the three specimens resemble *B. zanzibarica*, but the flowers are petalous; *Schlieben* 1577 has 5 or 6 sepals and petals, *Schlieben* 2078 has 4 or 5. *Haerdi* 385/0 has the normal 5 sepals and petals, but is in habit and nearly all other characters identical with *Semsei* 1327, a specimen without petals and doubtlessly belonging to 4, *B. zanzibarica*.

As in certain other annual species, sometimes large elliptic cotyledons are still present in full-grown annual specimens. In some annual forms bracts in the inflorescences are absent.

6. **B. lanceolata** (*Engl.*) *de Wilde* in Blumea **21**: 345, fig. 3/d–h (1973). Type: Tanzania, without precise locality, *Fischer* 268 (B, holo.†, BM, drawing of leaf)

Erect, climbing or prostrate annual or perennial, mostly branched, 15–80 cm., glabrous. Tendrils 2–10 cm., or absent. Leaf-blades unlobed or deeply (2–)3-lobed, lanceolate to broadly ovate, 1–12 by 0·5–4 cm., base subtruncate to narrowly cuneate or attenuate, top subobtuse to acute-acuminate; lobes elliptic to oblong; margin serrate-dentate with teeth up to 2 mm.; petiole (0·1–)0·5–5(–7) cm., usually narrowly to broadly winged and with small gland-teeth. Stipules 3–11 mm. False stipules absent. Inflorescences 1–2-flowered; peduncle (0·5–)1–4 cm.; bracts 1–3 mm. Flowers glabrous; stipe 3–8(–10) mm. Hypanthium 1·5–4 mm. wide. Sepals 5–11 mm., ± obtuse. Petals 2–7 mm. Outer corona-tube 1·5–2 mm., threads 1·5–2·5 mm. Disk 0·2–0·5 mm. Inner corona cup- or funnel-shaped, 0·6–2 mm. Stamens 5; filaments 2–3·5 mm.; anthers 1–4 mm. Ovary 0·6–2·5 mm.; styles free or up to 1 mm. connate, (1–)1·5–4 mm. Fruit (excluding the up to 5 mm. long gynophore) 1–1·8 cm., containing 1–5 seeds. Seeds ± 6 mm.

KENYA. Northern Frontier Province: Lolokwi Mt., 26 Dec. 1969, *Gillett* 18966!; Meru District: Meru Game Reserve, 11 Sept. 1963, *Verdcourt* 3749!; Kwale District: between Umba and Mwena Rivers, 18 Aug. 1953, *Drummond & Hemsley* 3856!

TANZANIA. Moshi District: Sanya Plains, Apr. 1967, *Beesley* 273!; Tanga District: Mtotohovu, 11 Sept. 1951, *Greenway* 8710!; Lindi District: Mbemkuru R., 15 Feb. 1935, *Schlieben* 5996!; Zanzibar I., *Herb. d'Alleizette* 2589! (in L, a deviating specimen)

DISTR. **K**1, 4, 7; **T**2, 3, 6–8; **Z**; not known elsewhere

HAB. Grassland and bushland, disturbed places; 0–1700 m.

SYN. *Tryphostemma lanceolatum* Engl. in E.J. **14**: 388 (1891); Harms in E.J. **15**: 577 (1893) & in P.O.A. C: 281 (1895); Hutch. & Pearce in K.B.: 264 (1921); Harms in N.B.G.B. **8**: 292 (1923)

T. cuneatum Engl., V.E. 3 (2): 598 (1921). Type: Tanzania, Tanga District, Moa, *Kassner* 46 (B, holo.†, BM, K, iso. !)

T. alatopetiolatum Harms in N.B.G.B. **13**: 424 (1936). Type: Tanzania, Lindi District, Mbemkuru R., *Schlieben* 5996 (B, holo.†, BM, BR, HBG, iso. !)

[*T. longifolium* sensu Agnew, U.K.W.F.: 166 (1974), *non* Harms]

NOTE. Very variable in leaf-shape. Decumbent specimens (e.g. *Renvoize 2215, Rawlins 133, 122*), as well as certain erect specimens (e.g. *Greenway & Kanuri 12810, Lenthold 105*, and *Schlieben 5996*) have long, narrowly winged petioles. Such specimens have relatively small flowers and fruits.

The perennial specimens *Bally 8000, 8001, 8361, Drummond & Hemsley 1263, Gilbert 4899* and *Verdcourt 3749*, from **K4** and **T2**, have lanceolate-linear leaves, sessile or with a very short unwinged petiole. In these specimens the tendrils are sometimes little developed or absent, and the leaf-margin is inconspicuously dentate.

A few specimens with some characters intermediate with *B. hanningtoniana* are discussed under that species.

The specimen *Herb. d'Alleizette 2589* (L), annotated as from Zanzibar, has a deviating habit. It consists of two shoots of ± 30 cm. from a woody rootstock. The leaves are lanceolate, with an unwinged petiole up to 2 mm., with 2 small tooth-like glands at the blade-base. The ovary is ± 1 mm. stiped (gynophore). The heterogeneity within this species, as well as in *B. hanningtoniana*, needs further study.

7. B. pubiflora *de Wilde* in Blumea 21: 347, fig. 4/b (1973).   Type: Tanzania, Mpanda District, Kapapa, *Richards* 11646 (K, holo.!)

Perennial with woody rootstock; shoots erect, unbranched, to 40 cm., finely pubescent. Tendrils absent. Leaf-blades unlobed, elliptic to elliptic-oblong, 2–4 by 1·2–2 cm., base obtuse to subacute, top acute, ± acuminate; margin subentire with a few hair-like glands less than 0·5 mm.; petiole 0·2–0·5 cm. Stipules 2–5 mm. False stipules absent. Inflorescences (1–)2(–3)-flowered; peduncle (0·5–)1–5 cm.; bracts 1·5–4·5 mm. Flowers pubescent; stipe 8–12 mm. Hypanthium 3–4(–5) mm. wide. Sepals (6–)8–11 mm., subobtuse. Petals 6–8 mm. Outer corona-tube 2–2·5 mm., threads (1–)2 mm. (variable in one flower). Disk 0·3–0·5 mm. Inner corona cup-shaped, 1–1·5 mm. Stamens 5; filaments 3–4·5 mm., anthers 3–3·5 mm. Ovary 2–2·5 mm.; styles 3–4·5 mm., connate for 1–1·5 mm. Fruit not known.

TANZANIA. Mpanda District: Kapapa Camp, track to Mpanda, 29 Oct. 1959, *Richards* 11646!; Iringa District: Ruaha National Park, Mangangwe Air Strip, 16 Dec. 1972, *Bjørnstad* 2130
DISTR. **T4, 7**; not known elsewhere
HAB. Deciduous woodland; 1050–1350 m.

NOTE. The Bjørnstad specimen was received on loan from Oslo after the description had been printed; it is somewhat more robust than the type.

8. B. hederae *de Wilde* in Blumea 21: 348, fig. 4/d (1973).   Type: Tanzania, Mpanda District, Kapapa, *Richards* 11613 (K, holo.!)

Perennial with woody rootstock; shoots prostrate, sparingly branched, to 35 cm. long; pubescent. Tendrils absent. Leaf-blades unlobed or shallowly 3(–5)-lobed in the upper half, suborbicular to reniform, 1–2·3 by 1·5–3 cm., base subtruncate to deeply cordate, top subobtuse to acute, margin entire with a few minute gland-teeth; petiole 1–4 mm. Stipules 1·2–3 mm. False stipules absent. Inflorescences 1–2-flowered, sessile; bracts ± 1·5 mm. Flowers pubescent; stipe 8–16 mm. Hypanthium 2–2·5 mm. wide. Sepals (5–)6–7 mm., subobtuse. Petals 3–5 mm. Outer corona-tube ± 1·5 mm., threads 3–3·5 mm. Disk ± 0·5 mm. Inner corona cup-shaped, 0·5–0·8 mm. Stamens 5; filaments 2·5–3 mm., anthers ± 2·5 mm. Ovary 1–1·3 mm.; styles ± 3·5 mm., free. Fruit not known.

TANZANIA. Mpanda District: Kapapa Camp, 28 Oct. 1959, *Richards* 11613!
DISTR. **T4**; known only from the type
HAB. Deciduous woodland; 1050 m.

NOTE. Known only from a single collection, from about the same locality as *B. pubiflora*.

9. **B. scabrifolia** (*Dandy*) *de Wilde* in Blumea 21 : 353, fig. 1/1 (1973). Type: E. Uganda, or W. Kenya, *V.G. van Someren* (K, holo.!)

Perennial climber to ± 3 m., scabrous; stem scabrescent. Tendrils absent or up to 7 cm. Leaf-blades unlobed, elliptic to oblong, 1·5–4·5 by 0·5–1·5 cm., base and top subobtuse to acute, margin ± 1 mm. serrate-dentate; petiole up to 0·5 cm. Stipules 1·5–4 mm. False stipules absent (sometimes small first leaves of serial shoot developed). Inflorescences 1–2-flowered; peduncle up to 4 cm.; bracts 2–3 mm. Flowers scabrous-hispid; stipe 8–12(–15) mm. Hypanthium 2·5–3 mm. wide. Sepals (6–)8–10 mm. Petals 4–6 mm. Outer corona-tube 2–2·5 mm.; threads 2·5–3 mm. Disk ± 0·5 mm. Inner corona cup- or funnel-shaped, ± 1·5 mm. Stamens 5; filaments 3–4 mm.; anthers 2–3 mm. Ovary ± 1·5 mm.; styles free, 3–4 mm. Fruit (excluding the 1–2 mm. long gynophore) 1·2–2 cm., containing 1–5 seeds. Seeds ± 7 mm.

UGANDA. E. Uganda, or W. Kenya, without locality, Apr. 1926, *V.G. van Someren*!
KENYA. Kisumu–Londiani District: Dunga–Kisumu, 13 Dec. 1968, *Kokwaro* 1689! & Kisumu, 3 Jan. 1969, *Kokwaro* 1807! & Sept. 1958, *Tweedie* 1686!
TANZANIA. Musoma District: Serengeti, Kirawira Plain, 27 Apr. 1965, *Richards* 20300!; Mbulu District: between Magugu and Babati, 7 May 1962, *Polhill & Paulo* 2368!
DISTR. ?U3; K5; T1, 2; not known elsewhere
HAB. Wooded grassland and deciduous bushland, sometimes in rocky places; 900–1400 m.

SYN. *Tryphostemma scabrifolium* Dandy in K.B.: 252 (1927)

NOTE. The specimen *Kokwaro* 1807 shows a woody main stem nearly 1 cm. across with a thick corky dark grey chapped bark.

10. **B. spinosa** *de Wilde* in Blumea 21 : 354, fig. 4/m (1973). Type: Kenya, Marsabit, *Bally* 5478 (K, holo.!, C, EA, iso.!)

Thorny shrub to ± 1 m., scabrous. Tendrils absent; thorns (1–)2–4(–5) cm. Leaf-blades elliptic to obovate, 0·5–2 by 0·2–1·5 cm., base subacute, top acuminate-mucronate; margin 0·5–1 mm. dentate; petiole 0·5–2 mm. Stipules 4–7 mm. False stipules absent. Inflorescences 2-flowered; flowers at or below the middle of the thorns, peduncular part of thorns 0·5–1·5 cm. Flower stipe 3–8 mm.; bracts (bracteoles) ± 3 mm. Hypanthium 2–3 mm. wide. Sepals 6–10 mm., obtuse, ± scabrous-hispid. Petals 4·5–6 mm. Outer corona-tube 1·5–2 mm., threads 1·5–2 mm. Disk 0·3–0·5 mm. Inner corona cup- to funnel-shaped, 0·7–1 mm. Stamens 5; filaments (1–)2–4 mm.; anthers 3–3·5 mm. Ovary 1·5–2·5 mm.; styles (1–)1·5 mm., connate for 0·2–0·5 mm. Fruit (excluding the ± 1 mm. long gynophore) 1–1·5 cm., containing 2–5 seeds. Seeds 5–6 mm.

KENYA. Northern Frontier Province: Marsabit to Lake Paradise, 4 Oct. 1947, *Bally* 5478! & Uaso Nyiro, Yaka, May 1945, *J. Adamson* 76 in *Bally* 4377! & Nyambeni Range, Locadema Hill, 20 Dec. 1971, *Bally & Smith* 14723!
DISTR. K1; not known elsewhere
HAB. Deciduous bushland, on lava or volcanic soils; 800–1300 m.

NOTE. The thorns, as in certain species of *Adenia*, are the modified peduncles and terminal (first) flower of the inflorescences; they replace the tendrils of other species·

7. SCHLECHTERINA

Harms in E.J. 33: 148 (1902) & in Ber. Deutsch. Bot. Ges. 24: 177, t. 12 (1906); V.E. 3(2): 596 (1921); Harms in E. & P. Pf., ed. 2, 21: 485, fig. 221 (1925); de Wilde in Blumea 22: 48 (1975)

Low perennial climber or suberect shrublet, usually provided with tendrils, glabrous, growing from a perennial rootstock. Leaves simple, elliptic to linear, entire to deeply pinnately lobed, shortly petiolate, margin entire to dentate. Glands on petiole 0–1(–2) pairs at the top. Stipules small, caducous. Tendrils axillary. Inflorescences axillary, sessile or subsessile, few-flowered, often arranged on short-shoots from the supra-axillary bud; bracts and bracteoles small. Flowers hermaphrodite, whitish; stipe articulate to short pedicel. Hypanthium small, shallowly cup-shaped. Sepals 3–4, free, imbricate, elliptic to oblong. Petals 2–4, free, elliptic to oblong. Corona single, composed of threads connate at base into a low tube, inside set with additional hair-like appendages. Disk absent. Androgynophore short; stamens 6–8, connate at base into a shallow cup, often with small lobes (staminodes) on its margin between the filaments; anthers dorsifixed, versatile, ellipsoid-oblong, obtuse, 2-thecous. Ovary ellipsoid-oblong, on a short gynophore; placentas (3–)4, each with 3–8 ovules; style single; stigma single, flattish, 3–4-lobed. Fruit a stipitate 3–4-valved capsule, ellipsoid-oblong, fusiform; valves coriaceous. Seeds flattened, ellipsoid, arillate; testa crustaceous, scrobiculate.

One species in tropical eastern Africa.

NOTE. The genus is related to the W. African genus *Crossostemma* Benth., and has been included by Hutchinson, G.F.P. 2: 370 (1967), but the latter is distinct by having a conspicuous intra-staminal disk and no gynophore.

S. mitostemmatoides *Harms* in E.J. 33: 148 (1902) & in Ber. Deutsch. Bot. Ges. 24: 177, text fig., t. 12 (1906); Bak. in J.L.S. 40: 73 (1911); Engl., V.E. 3(2): 596 (1921); Harms in E. & P. Pf., ed. 2, 21: 485, fig. 221 (1925); A. & R. Fernandes in Garcia de Orta 6: 249 (1958). Type: Mozambique, Lourenço Marques, *Schlechter* 11681 (B, holo.!)

Small liana or subscandent shrublet to 3 m., glabrous, with perennial rootstock; older stem corky, shoots often lenticellate. Tendrils 3–10(–15) cm. Leaf-blades variable (heterophyllous), elliptic to lanceolate or linear-lanceolate, top acute to long-acuminate, base acute to attenuate, margin entire or regularly to irregularly dentate, or leaves pinnately lobed to various depth, the lobes acute to rounded, 2·5–13 by 0·7–4·5 cm.; leaves on saplings, sterile or juvenile shoots often lanceolate to linear, ± deeply pinnately lobed, the lobes broad or narrow, 10–30 by 0·2–2 cm.; petioles 4–12 mm., on juvenile plants 0–4 mm.; glands on petiole absent or 1(–2) pairs at the top; glands on blade-margin several, minute, mostly at the tips of the teeth; stipules subtriangular to linear, ± 0·5 mm. Inflorescences 1–3-flowered; bracts subtriangular, 0·5–1 mm. Flowers glabrous; stipe 6–25 mm. Hypanthium shallowly cup-shaped, ± 3–4·5 mm. wide. Sepals elliptic to oblong, obtuse, 6–11 by 3–6 mm. Petals elliptic to oblong, obtuse, 5–10 mm. long. Corona 5–8 mm. high, composed of threads ± connate at base into a tube 0·5–2 mm., free parts of threads 4–6 mm., inside tube and at base of free threads a zone of short hair-like appendages 0·5–1 mm. Androgynophore ± 1 mm., filaments 6–10 mm., connate at base into a cup ± 1 mm., often with small lobes on its margin in between the filaments; anthers 2–3 mm. Gynophore 2–2·5 mm.; ovary 2–2·5 mm., ± (3–)4-angled; style 1·5–2 mm.; stigma 2–2·5 mm. across. Fruit ellipsoid-oblong, acute (fusiform), excluding the ± 1 cm. long gynophore 4·5–5 by 2·5–3 cm. Seeds rather few, ellipsoid, ± 8 mm. Fig. 9.

KENYA. Kwale District: Shimba Hills, Lango ya Mwagandi [Longo Mwagandi], 13 Mar. 1968, *Magogo & Glover* 287; Kilifi District: Marafa, Jan. 1937, *Dale* in *F.D.* 3648!; Lamu District: Witu, June 1957, *Rawlins* in *E.A.H.* 11269!

TANZANIA. Tanga District: 8 km. SE. of Ngomeni, 31 July 1953, *Drummond & Hemsley* 3577!; Uzaramo District: Pugu Hills, 17 Sept. 1967, *Harris* 914!; Newala District:

Fig. 9. *SCHLECHTERINA MITOSTEMMATOIDES*—**1,** habit, × ⅔; **2,** flower, × 4; **3,** stamens, × 6; **4,** pistil, × 8; **5,** fruit, × ⅔; **6,** seed, × 3. 1, from *Faulkner* 1773 & *R.M. Graham* B.467; 2–4, from *Padwa* 791; 5, 6, from *Hornby* 656. Drawn by Victoria Goaman.

Mahuta, Dec. 1942, *Gillman* 1066!; Zanzibar I., Makunduchi, 16 July 1933, *Vaughan* 2139!

DISTR. **K**7; **T**3, 6, 8; **Z**; Mozambique and South Africa (N. Natal)

HAB. Lowland dry evergreen and riverine forest, coastal bushland; 0–700 m.

SYN. *S. mitostemmatoides* Harms var. *holtzii* Harms in Ber. Deutsch. Bot. Ges. 24: 184 (1906); V.E. 3 (2): 596 (1921); Harms in E. & P.Pf., ed. 2, 21: 486 (1925). Types: Tanzania, near Dar es Salaam, *Holtz* 1070 & 1070a (B, syn.!)

NOTE. The species displays a remarkable variability in leaf-shape.

8. EFULENSIA

C.H. Wright in Hook., Ic. Pl. 6, t. 2518 (1897); G.F.P. 2: 373 (1967); de Wilde in Blumea 22: 31, 48 (1975)

Deidamia sensu Harms in E. & P. Pf. III. 6A, Nachtr. 1: 254 (1897) & in E.J. 26: 239 (1899)

Georgiella De Wild. in F.R. 13: 384 (1914)

Deidamia Thouars sect. *Efulensia* (C.H. Wright) Harms in E. & P. Pf., ed. 2, 21: 487 (1925)

Large climbers, glabrous. Tendrils simple, axillary or replacing the central flower of the inflorescences. Leaves 3- or 5-foliolate; petiole distinct; leaflets obovate or elliptic to oblong, entire, distinctly petiolulate. Glands on petiole 1 pair, subopposite, situated towards the base; on leaflets marginal or submarginal, small, or absent. Inflorescences axillary, cymose, 2–100-flowered; peduncle distinct. Flowers hermaphrodite or functionally ♂, glabrous. Stipe distinct, jointed to short pedicel. Hypanthium saucer-shaped, 2–3 mm. wide. Tepals reflexed in anthesis; sepals 5, free; petals 5, free. Corona single, 3·5–8 mm. high, composed of threads connate at base into a short fleshy tube; threads towards base inside with a zone of short hair-like appendages ± 0·5 mm. Androgynophore ± 0·5 mm. Stamens 5; filaments 3–10 mm., united at base into a broad cup 1–1·5 mm., in *E. clematoides* sometimes with 5 small teeth at the bottom of the cup and alternating with the stamens; anthers dorsifixed, versatile, ellipsoid (to oblong), apiculate or not. Disk, besides small teeth, absent. Gynophore 0·5–1 mm.; ovary subglobose to ellipsoid, 1-locular, with 3(–4) parietal placentas. Styles 3(–4), free or up to over half-way connate; stigmas sub-globose, glabrous, 1–1·5 mm. across. Fruits 1–6 per inflorescence, capsular, woody, 3-(4)-valved, subglobose, depressed globose or ellipsoid, 1·5–4 cm. long. Seeds 4–12 per capsule, subellipsoid, 6–8 mm., coarsely pitted.

A genus with 2 species in equatorial Africa, from S. Nigeria to W. Uganda.

Efulensia was formerly included by several authors in *Deidamia* Thouars from Madagascar, which it resembles very much in habit and by the 3- or 5-foliolate leaves. *Deidamia* differs by the 5–8 stamens free to the base, and by the presence of a distinct annular extrastaminal disk.

E. montana *de Wilde* in Blumea 22: 34, fig. 1/c, 47, fig. 3/c (1975). Type: Zaire, Mushweri–l'Vrega, *Lebrun* 5574 (BR, holo.!)

Liana to 20 m. Tendrils (including peduncle) 8–20(–25) cm. Leaves 5-foliolate, palmate to imparipinnate, the lowest pair of leaflets inserted at a distance of up to 3 mm. from top leaflets; leaflets elliptic to oblong, top mostly acute, up to 0·5 cm. acuminate, rarely subobtuse, up to 1 mm. mucronate, base subobtuse to acute, (1·5–)2·5–10 by (0·7–)1·2–3·5 cm.; petiole (1·5–)2–8 cm.; petiolules 0·3–1·2 cm., when dry not jointed to the petiole. Glands on petiole 0·5–2 mm. across, situated $\frac{1}{5}$–$\frac{1}{3}$ (i.e. 0·5–2·5 cm.)

FIG. 10. *EFULENSIA MONTANA*—**1,** flowering branch, × ⅔; **2,** flower, × 4; **3,** stamens, × 6; **4,** ovary, × 6. All from *Purseglove* 2652. Drawn by Victoria Goaman.

from the base of the petiole; glands on leaflets (0–)1(–2), ± 0·5 mm. across, at each side on the margin close to the base. Inflorescences 2–6(–8)-flowered; peduncle 5–13 cm. Flowers hermaphrodite or functionally ♂. Stipe 7–25 mm. Sepals 7–10 by 2·5–3 mm., obtuse. Petals 7–9 by 2–2·5 mm., acute to obtuse. Corona 3·5–6 mm.; threads 2–4 mm., tube 1·5–2 mm. Filaments 3–6 mm., united at base into a broad cup ± 1 mm.; anthers 2·5–3·5 mm., ± 1 mm. apiculate or not. Ovary in ♂ flowers much reduced, 1·5–2 mm.; in hermaphrodite flowers ellipsoid, 3–3·5 by ± 1·5 mm.; styles 3, ± 1·5 mm., connate for 0·5–1 mm. (free style-arms 0·5–1 mm.). Fruits 1–4 per inflorescence, ellipsoid, obtuse to subacute at both ends, excluding the 0·5–1 cm. long gynophore 2·2–4 by 1·5–2·3 cm.; valves 2·5–3 mm. thick at sutures. Seeds 6–12 per capsule, ± 6 mm., with 3–4 pits across the length. Fig. 10.

UGANDA. Kigezi District: Kanungu, Oct. 1940, *Eggeling* 4193A! & Kayonza, Apr. 1948, *Purseglove* 2652!
DISTR. **U**2; E. Zaire
HAB. Forest; 900–2000 m.

EXCLUDED GENUS

Donaldsonia Bak. f. in J.B. 34: 53, *s*. 355/A (1896).

The single species of *Donaldsonia*, from the NE. corner of Lake Rudolf, belongs to *Moringa* as *M stenopetala* (Bak. f.) Cufod. in Senck. Biol. 38: 407 (1957).

INDEX TO PASSIFLORACEAE

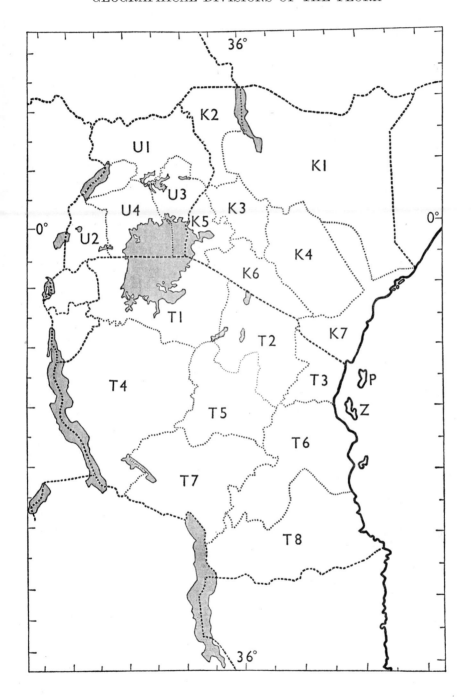